Parry's
Graining and Marbling

PARRY'S
GRAINING
AND
MARBLING

John P. Parry

Second Edition

Revised by
Brian Rhodes and John Windsor

COLLINS
8 Grafton Street, London W1

Collins Professional and Technical Books
William Collins Sons & Co. Ltd
8 Grafton Street, London W1X 3LA

First published in Great Britain by
Crosby Lockwood & Son, Ltd 1949
Second Edition published by
Collins Professional and Technical Books 1985
Reprinted 1985, 1986, 1987

Distributed in the United States of America
by Sheridan House, Inc.

British Library Cataloguing in Publication Data
Parry, John P.
Parry's Graining and marbling.—2nd ed.
1. Graining
I. Title   II. Rhodes, Brian   III. Windsor, John.
745.51   TT330

ISBN 0–00–383131–0

Typeset by V & M Graphics Ltd, Aylesbury, Bucks
Printed and bound in Great Britain by
Mackays of Chatham, Kent

Contents

List of Plates

Foreword to First Edition
by James Lawrance, F.I.B.D.

We should have to search far back into the records of ancient history to discover the origin of imitative painting as applied to woods and marbles. Perhaps in its earlier stages the craft was little more than an attempt craftily to counterfeit foreign woods and marbles that were not easy to obtain. William Wall, in his book *Graining Ancient and Modern*, calls our attention to a passage in a work entitled *Museum of Antiquity*, published in 1882, which claimed to be a description of ancient life three thousand years ago. He quotes the authors thus :

> 'Boxes, chairs, tables, sofas, etc., were often made of ebony inlaid with ivory, sycamore, and acacia veneers, with layers and carved devices of rare wood added as ornament on inferior surfaces; and fondness for display induced the Egyptians to paint common boards to imitate foreign varieties, so generally practised in other countries at the present day. The colours were usually applied on a thin coating of stucco or a ground smoothly laid on prepared wood, and the various knots and grains made to resemble the wood they intended to counterfeit.'

William Wall comments: 'This account would appear to indicate that grainers were a professional class of artisans over three thousand years ago,' and tells us that 'there is shown in the British Museum in London a bill of account several centuries old, for painting and graining a room in the Tower of London.'

From this 'fondness for display' sprang a craftsmanship that developed into something greater than mere imitative painting, and the examples in this book, executed by Mr Parry, express the richness of this craftsmanship better than any words of mine could convey.

In 1830 Messrs W. and T. J. Towers wrote a book on Painting and Graining, and it is interesting for present followers of the craft to read the instructions there given for 'imitating oak in oil for outside work' as follows:

'Mix rotten stone and white lead to the tint required, for the ground: when this is dry, and made perfectly smooth with fine glass-paper the graining colour may be applied; for which take four ounces of rotten stone and four ounces of sugar of lead and grind them quite stiff in boiled linseed oil. ... Put a small quantity of "grainer's cream" (bee's wax melted and thinned with turpentine) into the graining colour in order to keep it from flowing together. ... This style of work requires working with combs of various sizes; the best are made of ivory; but if these cannot be obtained; either wood or leathern ones may be substituted.

Spread the colours over the surface of the work with a large paint brush, about half worn, take a coarse comb and pass over it in a straight direction, pressing moderately hard after which, take a finer comb and pass over several times in a wavy direction; then, with an ivory tooth comb, with the two outside teeth broken off, pass over the centre of the work with a very tremulous motion of the hand, in order to produce the finest grain which is in the centre of the tree. To produce the flower or veins, use a piece of thin wash leather wrapped tightly round the thumb.

When the whole of your work is dry, dip the flat hog's hair graining brush into a small quantity of burnt umber, ground up in ale, very thin, and pass over it in a straight direction: this will leave the fine transparent grain, so natural to this beautiful wood. When dry, varnish.'

A comparison between this and Mr Parry's instruction, descriptive of present methods, serves to mark the progress that has been made in developing a craft that can claim the respect due to all continuous and progressive exercises of manipulative skill when their origins are so deeply rooted in antiquity.

Woods and marbles provide fine motifs as a playground for the craftsman. They give scope for the expression of those subtleties of light and shade, of colour and of translucent gradation that rarely find such abundant display as in woods and marbles.

The suggestion that graining and marbling is inadmissible because it is imitative rests on very slender grounds. There is perhaps no interpretative effort that can so smoothly flow from the tools of the craftsman without any attempt at slavish imitation. Graining and marbling find scope for the craftsman for exerting the same kind of control over his tools as he does over his limbs, as if the tool were grafted on to his body and responded promptly and almost automatically to this control. The markings of woods hold all those creative curves and repetitions that are most pleasing to the eye, and thus give full scope for interesting interpretation; and are not marbles the result of the great accidents of Nature, thus lending ideal motifs for the accidental yet skilful and ingenious methods of opening varied pockets in a sea of colour?

Graining and marbling have been executed for centuries, and the insistent, continuous demand for such work is, in itself, a good ground for retention of a craft practice which aims at the suitable and correct interpretation of wood and marble motifs. The serviceable character of this work, arising from the broken-colour system peculiar to it, has fully established its claim, and has been an essential factor in creating this demand.

Few grainers would agree that their renderings were imitative, and any such claims would usually prove very unconvincing. They obtain more satisfaction from their ability to display in paint those characteristics which suggest the 'feeling' of wood and marble, thus expressing their own appreciation of the beautiful structure of natural growth or natural calamity as the case may be.

Mr Parry clearly defines, by his examples as well as by his words, the acceptable scope for the efforts of the grainer – to express the beautiful character of the woods and marbles rather than to blunder into ineffectual imitation. I commend his chapters to readers with full confidence that they will stimulate an appreciation of the correct approach to the subject as well as the right kind of craftsmanship for implementing the conception thus promoted. Mr Parry can do all this because he is master of his subject.

Preface to the Second Edition
by Brian Rhodes and John Windsor
1985

Even though there appeared to be no text book available to meet the growing resurgence of graining and marbling, it was never our intention to write a new book on the subject. We have of course developed our own ideas and techniques during our professional careers, but *Parry's Graining and Marbling* has for many years been established as the most comprehensive book on the subject, and we would not wish to compete with it.

Unfortunately the book has long been out of print, and we decided a second edition revival would meet the demand for a contemporary work on the subject.

The style and content of Parry's work has been left unchanged. All we have done is to review and update the materials in the light of changing technology and availability, and to introduce BS 4800 in an attempt to standardise the ground colours suggested by Parry for the various woods.

Some of the woods described by Parry are no longer commonly available, and samples may be difficult to obtain. However, his descriptions of graining, and the colours used can be applied to the many woods which have now replaced them, and which fall into the same category or type.

The addition of explanatory sketches within the text is designed to supplement Parry's rich and descriptive vocabulary of the various techniques of graining and marbling, and to illustrate the use of the various tools and appliances.

We have enlarged, and hopefully enhanced the book with a chapter on the various broken colour effects which also involve the tools, materials and skills of graining and marbling. This additional work makes the book more comprehensive, more appealing to the special needs of the younger trainee studying at Advanced Craft level, and to all those who are either specifying or executing these popular decorative effects.

Finally, our thanks go to Susan Rhodes who kept 'badgering' us to achieve the agreed deadline for the revised work, and who kindly typed the manuscript.

Chapter One

Decorative possibilities of graining and marbling

Little is known concerning the true origin of these crafts, but there is sufficient evidence to support the belief that these specialised branches of painting were developed in England during the eighteenth century. Early examples were often crude until the 1850s, when that genius Thomas Kershaw of Westhoughton near Bolton, and his contemporary John Taylor of Birmingham, achieved a degree of perfection which to this day is seldom equalled. Their fame spread to the Continent, where, at a Paris exhibition, it required a practical demonstration to convince sceptical critics that the works shown were painted representations and not panels of the real woods and marbles.

It is a fitting tribute that the work of these excellent craftsmen has been carefully preserved and may still be viewed by students or others interested. Panels executed by Kershaw are exhibited in the Chadwick Museum, Queen's Park, Bolton, and the Victoria & Albert Museum, Kensington, London.

Graining, in common with gilding, veneering, the use of fibrous plaster, tile and other effects in wall-paper, relief decorations, manufactured marbles and other materials and processes, has long been accepted and appreciated as a traditional form of decoration: they do not attempt to deceive the observer, but rather to convey to the mind, by suggestion, the abstract idea expressed. Their utilitarian advantages are fully recognised, but their aesthetic value depends entirely on the exercise of good taste in selecting the most fitting material, pattern, or finish for a specific surface.

Perhaps the widest appeal of good graining lies in its unobtrusive richness, quality of finish, variety of colouring and pattern, and the soft translucency enhanced by subtle effects of light and shade. These qualities, displayed in a rather conventional manner (except in special cases) and in the right positions, possess greater decorative value than the slavish detailed copying of poor examples of the wood concerned.

Graining and marbling offer wide possibilities, as yet only partly explored. They need not, as in the past, be limited to realistic or impressionistic painting, but may well be taken a stage farther and conventionalised in both design and colour to the point of extreme simplification. Rag rolling, rag stippling and scumbling are examples of this idea which originated from marbling and which frequently offer a more colourful means of completing a harmonious scheme.

The introduction of plastic and other new materials in buildings will undoubtedly call for greater imagination and considerable modification in the treatment of woodwork. It will be for the grainer-craftsmen of the day to decide how far to sacrifice realism to the interests of unity, colour harmony, and fitness. This line of thought suggests an increasingly wider use of the light-toned wood effects, such as maple, pine, ash and limed oak, and of egg-shell gloss and dead flat finishes in oil and cellulose varnishes.

Where and where not to employ graining and marbling was decided long ago by such eminent authorities as Owen Jones, Sir Digby Wyatt, and others who condemned the use of painted representations in all positions where the real materials would be out of place. In this respect good taste is based upon sound common sense, and painters must remember this if they are to avoid such incongrous combinations as a marbled dado with grained plinth, marbled mouldings upon a grained door or the graining of iron railings and rain-water pipes. It is unfortunate that these and similar atrocities have at times been committed.

Chapter Two

The preparation and painting of surfaces

The intention of this chapter is to stress the importance of a high quality painted surface on which to work, and methods of achieving the standard required. It does not specify particular treatments for the preparation and painting of the many surfaces commonly used, for even though this specialist information is of great importance, its technical nature is outside the main purposes of this book, which is to describe the methods and materials used to produce good graining and marbling.

For specifications on the painting of surfaces a suitable textbook is *Painting and Decorating: An Information Manual*, published by Granada, and written by PADIM Technical Authors. Less detailed, but adequate information is produced by the major paint manufacturers, and is usually available free of charge upon request.

For both graining and marbling to deceive the eye and appear authentic, it is absolutely essential that the ground colour upon which the work is carried out is smooth, level and above all, free from brushmarks. The effect of highly polished marble, or the flowing patterns of wood grains are immediately diminished if attempted on a surface containing brushmarks and specks of dust, so the ground colour must be applied to a perfectly smooth surface without evidence of these defects, if the overall effect is to appear realistic.

Previously painted surfaces

These must be in sound condition, i.e. no flaking, blistering or peeling. They require washing, and wet abrading with waterproof silicon carbide paper (wet or dry). Use the abrasive paper wrapped around a block of rubber or felt to maintain a flat surface, and use small loose pieces of abrasive paper on mouldings.

Various grades of abrasive paper are available, and are selected according to the condition of the surface to be abraded. Surfaces containing coarse brushmarks, and paint runs initially require a strong grade of 150 to 240 to remove the surface defects. This will however scratch the surface and the scratches should then be removed by further abrasion with a finer grade of 320 to 400. Finally thoroughly rinse the surface with clean water and allow to dry. Painted surfaces in better condition may only require wet abrading with the finishing grade of abrasive paper, i.e. 320 to 400.

Any surface defects or irregularities should be 'made good' or repaired. Deep holes and cracks will first require 'stopping' to a level just below the surface, and then 'filling' with a fine surface synthetic filler, or an oil-based 'spachel' type filler. Small irregularities only require filling. Finally, abrade using whenever possible a rubbing block to support the abrasive paper to achieve a smooth, flat surface.

New surfaces

The nature of new surfaces which may receive paint are varied. They can be smooth, textured, absorbent, non-absorbent, chemically active or corrodible. Because of this, each requires a specialist pre-treatment and priming paint. Primers are specially formulated to suit the nature of the surface, and to provide a sound foundation for further coats of paint. Reference to the textbook mentioned earlier will provide the necessary information to enable the correct choice of appropriate pre-treatment and priming paint for any particular surface.

It is important to remember that paint alone will not fill surface irregularities, and if the new surface possesses texture, it will require 'face filling' all over with a fine surface filler, or an oil based 'spachel' type filler before further coats of paint are applied.

Mixing the ground colour

Alkyd eggshell finish paint is recommended for the ground of all graining, marbling and other broken colour effects. It offers a smooth non-absorbent surface without being shiny and slippery and supplies adhesion for the graining colour applied on top.

BS 4800

British Standard 4800 'Paint Colours for Building Purposes' is a colour card of a hundred colours used by the majority of paint manufacturers for matching purposes. This ensures that irrespective of the manufacturer, or the current fashionable names for colours, paints bearing the same BS 4800 code are almost identical, and available from most paint stockists.

The following chart contains a list of types of woods and a suggested ground colour for each. The closest BS 4800 colour has been specified, but these can be adjusted to suit the tone of the work required. A 'tint' of a colour has white added to make it lighter. A 'shade' of a colour has black added to make it darker. The ground colour can also be made by adding oil stainers to eggshell finish paint, and the chart contains a suggestion of the pigments which can be used to mix the colour. The first named forms the base of the mixture, with the other pigments successively reduced in quantity. Pigments should be ground on a palette board with a little of the base colour until mixed, and then added to the paint and thoroughly stirred in. Always strain the mixed colour before use.

Ground colours

Type of wood	Closest BS 4800 colour	Pigments used
Ash	Tint of 08.C.35	White, ochre, raw umber.
Light Oak	Tint of 08.C.35	White, ochre, raw umber.
Limed Oak	08.B.17	White, raw umber, burnt sienna.
Mahogany	04.D.44	Venetion red, ochre, white.
Maple	08.C.31	White, ochre.
Medium Oak	08.C.35	White, ochre, raw umber.
Pine	08.C.31	White, ochre.
Pitch Pine	06.C.33	White, yellow chrome, venetian red.
Pollard Oak	08.C.35	White, ochre, raw umber.
Rosewood	06.E.56	Orange chrome, venetian red, white.
Satinwood	10.C.31	White, yellow chrome.
Teak	08.C.35	White, ochre, raw umber.
Walnut	06.C.37	Ochre, burnt sienna, burnt umber.
Weathered Oak	00.A.01	White, black.

Note: Where ochre, umber and venetian red are used, increase the amount of liquid driers added to ensure overnight drying.

Applying the ground colour

A method of achieving the high quality finish required is to apply two or three coats of coloured eggshell finish paint. When thoroughly dry and hardened, wet abrade as previously described to remove any brush marks or specks of dirt which may contaminate the surface. Then apply the final coat of ground colour according to the manufacturer's instructions. Avoid brushmarks and bittiness by following the list of precautions described on page 106–7.

Plate 1

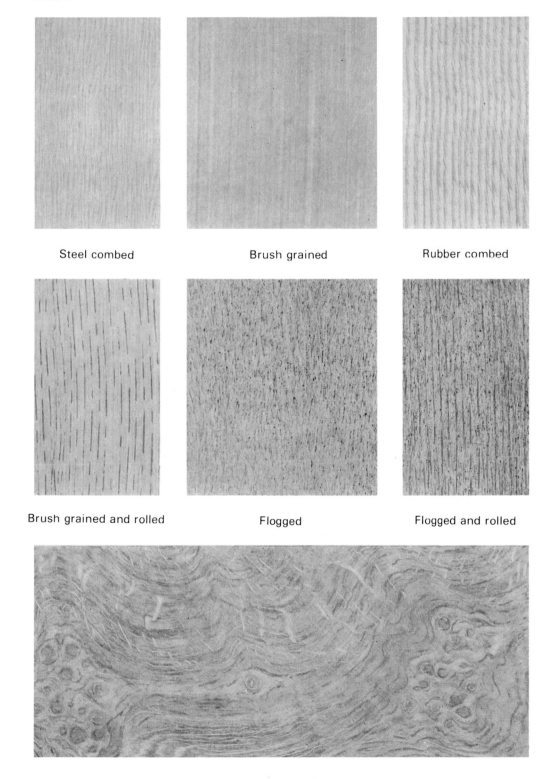

Steel combed

Brush grained

Rubber combed

Brush grained and rolled

Flogged

Flogged and rolled

Pollard oak

Plate 2

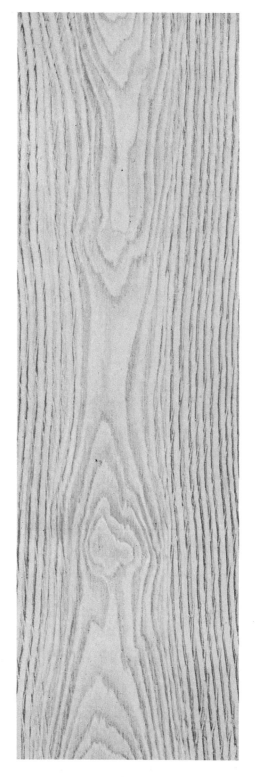

Oak (heartwood)   Quartered oak

Plate 3

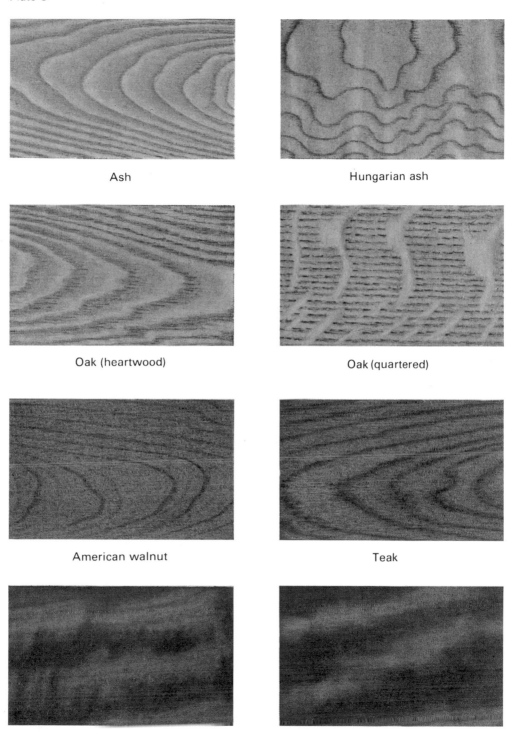

Ash

Hungarian ash

Oak (heartwood)

Oak (quartered)

American walnut

Teak

Mahogany

Old mahogany

Plate 4

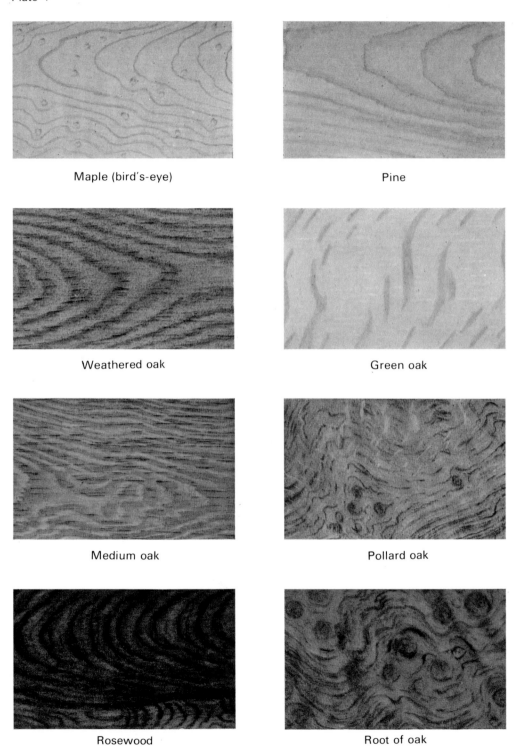

Maple (bird's-eye)

Pine

Weathered oak

Green oak

Medium oak

Pollard oak

Rosewood

Root of oak

Chapter Three

Graining colours: their uses and preparation in various media

The various graining colours whether in oil or water media, are all prepared from one group of pigments which, although limited in number, possess the valuable characteristic of translucency. It is this property which, when sufficiently reduced by a medium, produces the clean, semi-transparent glaze colours so essential in all forms of staining, scumbling, marbling, and graining.

It is a matter of first importance to use only the best of pigments, for these will be finely ground and unadulterated with material of a coarser and more opaque nature. It is difficult to obtain the required degree of cleanliness when working with cheap stainers. Tube colours and semi-prepared scumble glazes are not uneconomical as bases for oil-colour use; the amount required is almost negligible when compared with the remarkable spreading capacity of the graining colour. Furthermore, the standard of purity and colour remains fairly constant.

Pigments

The following list of pigments forms a palette from which graining colours may be mixed:

Raw umber; burnt umber; raw sienna; burnt sienna; Vandyke brown; mahogany lake; ivory drop black; Prussian blue.

These are available as dry colours and also as finely ground paste in oil, or water.

Paste colours purchased in small tubes or containers are preferable to and less wasteful than large containers. Small quantities are quickly used up and have little opportunity of losing their qualities of cleanliness and smoothness. Oil colours in tubes remain in excellent condition over a long period, but half-used tins of colour depreciate rapidly unless air is excluded by a protective covering of linseed oil or white spirit. Paste colours in water will naturally require to be kept moist by frequent additions of water; therefore it is essential that rustless containers be employed.

The media

As has already been indicated, both oil and water mediums are employed in graining. Each has peculiar advantages and limitations, and these must be fully explored in the interests of efficiency, economy, and sound technique. It will soon be apparent that with increasing command over tools and materials the scope for individuality and style is proportionately enlarged until perfection of technique results in what may now seem impossible – the combination of ease, speed, and quality.

Water graining colour

Dries rapidly and accumulates no dust; the coating is extremely thin, and for this reason facilitates the maintenance of surface flatness and smoothness; water stain is not liable to darken as a result of oxidation; its transparency is excellent; effects of extreme softness or hardness are easily obtained; work grained can be varnished within an hour. There are, however, certain limitations to be noted: the medium dries too rapidly to permit its use upon large unbroken areas; it is not easy to maintain perfect evenness of colour at the ends of moulded panels; medium and dark tones require two coats of varnish, but the latter has definite compensations in the form of increased durability and gloss.

Oil graining colour

Allows ample time for manipulation and is equally suitable for large and small areas; it offers the best medium for the combing and figuring which characterise oak; the work is usually finished with one coat of varnish; oil colour may be regulated to suit climatic conditions and can thus be evenly and easily applied in hot weather; mottled and softened effects are quite possible when the operator has acquired the necessary experience or technique, and when this standard is attained the grainer will be able to dispense with much of the overgraining hitherto necessary.

Alternate coatings of oil and water stain are frequently employed in the graining of walnut and other woods. This combination enables a job to be grained in water, fixed and overgrained in oil stain, and, when dry, glazed in water and finished with one coat of varnish.

Crayons

These provide yet another means whereby time can be saved and the quality of workmanship vastly improved. Again, we are not restricted to one medium, for crayons may be used upon and softened into a wet oil stain or – as in the graining of maple – they can be employed upon a dry coating of water stain to produce an effect quite free from the tone contrasts associated with brush-applied pattern.

The preparation of oil graining colour

The importance of cleanliness has already been stressed; it will therefore be taken for granted that paint kettles and brushes should likewise be clean. Our first job is the preparation of the medium or gilp in quantity sufficient for the work in hand, and for this purpose it is fairly safe to assume that 500 ml of medium will cover 25 square metres of surface.

The proportions of oil and turpentine must be varied according to the type of pigment employed; an oil scumble, for example, will require about three parts turpentine to one part of linseed oil, with a little liquid oil driers. With pigments ground in oil we shall require two parts turpentine to one part of boiled linseed oil and enough driers to ensure quick setting: driers vary in strength, but an amount equal to one-quarter of the weight of oil is usually sufficient.

Mixing is best carried out with the aid of an old brush, and, following the thorough breaking up of driers, the medium is ready for staining. The accompanying table gives the colours employed for various woods, irrespective of the type of medium used. Colours are added, a little at a time, and well stirred until thoroughly incorporated. By this method we gradually deepen the colour until (by repeated trial) the correct tone is obtained: excessive contrast is easily avoided and we prepare just the amount required for the job.

The following chart contains a list of types of wood and a suggestion of the pigments used to make the graining colour. The first named forms the bulk of the colour with the other pigments successively reduced in quantity.

Graining colours

Type of wood	Pigments used
Ash	Raw umber, raw sienna, black.
Light Oak	Raw umber, raw sienna, black.
Limed Oak	White, raw umber, black.
Mahogany	Vandyke brown, mahogany lake or burnt sienna.
Maple	Raw umber, raw sienna.
Medium Oak	Burnt umber, raw sienna, black.
Pine	Raw umber, raw sienna, burnt sienna, black.
Pitch Pine	Raw sienna, burnt sienna, burnt umber.
Pollard Oak	Burnt umber, raw sienna, burnt sienna.
Rosewood	Vandyke brown, mahogany lake, black.
Satinwood	Raw sienna, burnt sienna, Vandyke brown.
Teak	Burnt umber, burnt sienna.
Walnut	Vandyke brown, black.
Weathered Oak	Black, burnt umber, ultramarine blue.

Prepared scumble

Semi-prepared oil scumble glazes, or graining colours are readily available in a range of colours suitable to imitate most common woods. They can also be intermixed or further coloured with oil stainer to enable many other woods to be imitated.

They only require stirring, and thinning with turpentine before being applied to a suitable ground colour. The manufacturers supply a range of ground colours, or undercoats to suit each of the prepared scumbles.

After application, prepared scumbles are touch dry in about two hours, and dry to a matt finish. The dull surface is ideal to receive overgraining in water graining colour without undue cissing.

Megilp

This is a substance used by grainers who experience some difficulty in making their work 'hold up' or retain its sharpness of definition after combing, etc. Among the strange assortment of megilps favoured by past generations of grainers are: limewater, rainwater, jellied soap, whiting, and beeswax, the last-named being dissolved in hot linseed oil to facilitate mixture with the medium.

There are potential dangers connected with the use of any one of these: the beeswax, in strict moderation, is probably the safest of this risky group, but an excessive quantity of any megilp thickens the graining colour, gives an unpleasant relief effect, and produces unnatural sharpness of contrast.

Among the more recent innovations are several proprietary materials described as scumble glaze, scumble medium, etc. These are transparent compositions possessing considerable viscosity; they mix readily with oil and turpentine, and are, in effect, present-day megilps of an improved type.

Graining colour can be made to 'hold up' without these aids; it is simply a matter of balancing the ingredients so as to produce a rather 'sharp', quick-setting colour, which is well brushed out during the 'rubbing-in' process. It will also be found that finely ground paste driers and oil-scumble glazes serve quite well as supporting agents.

The preparation of water graining colour

These consist of pigment, a binder or fixative, and water for thinning purposes. The binders in general use are stale beer, fuller's earth, vinegar, and skimmed milk: any one of these may be ground with the paste water colour and subsequently diluted to the consistency required.

The first operation calls for a palette knife and a flat clean surface; after the preliminary grinding and slight thinning the pigment is mixed – as in oil-colour work – with the appropriate amount of thinner, prepared as follows:

(1) Binder of beer or vinegar; thinned with three parts of water.
(2) Binder of fuller's earth; thinned with water only.
(3) Binder of skimmed milk; thinned with equal parts of milk and water.

Care must be taken when using the binders listed in group (1), for these, if too strong, may cause the graining to crack; furthermore, the acidity of these mixtures must have a damaging effect upon superimposed varnish, and this necessitates two coats, the first of which may be thinned with turpentine.

A small amount of fuller's earth forms an effective and harmless fixative which also eliminates 'cissing' and thus ensures easier and speedier application. Vandyke brown and burnt sienna may be used without the addition of a binding agent when employed in the form of a thin wash or glaze.

Water graining colour is reversible when dry. This means that it can be softened with water, or a further application of water graining colour.

Because experienced grainers are familiar with the pattern, or figuring they are about to imitate, and are very proficient in the manipulation of the graining colour, they are able to apply successive coats of water graining colour without disturbing the dried coating underneath. However, the not so skilled grainer should always 'strap down' or 'fix' each coat of colour with a mixture of equal parts of eggshell varnish and white spirit. When dry, the fixed surface can be 'degreased', and further coatings of water graining colour applied without disturbing the one underneath.

All water stains require frequent stirring to maintain an even colour during application, and, where the drying speed is uncomfortably rapid, a few drops of glycerine will serve as an excellent retarder, but any excess in the amount used may well prevent thorough hardening.

Crayons

Although the idea is by no means new, there are many grainers who have never explored the wide possibilities of crayon work. Here is an extremely versatile medium peculiarly adaptable to certain processes in graining and marbling. It can be used upon and softened into the wet oil stains when representing oak, teak, and other heartwood figures, and is equally suitable for sketching in the finer lines upon a dry ground of water stain: it makes possible the combination of ease and speed with a sketchy quality quite distinct from that produced by brush application.

Among the ready-made crayons, those manufactured by Conté of Paris are suitable for oil-colour work, but crayon pencils of the appropriate colour are preferable for water-colour graining; the latter should always be tested to ascertain their behaviour when varnished, for there are just a few varieties which are inclined to run.

Chapter Four

Brushes, tools and appliances

As this chapter deals with a complete grainer's kit, we shall overlook the fact that no two grainers think alike on this subject. Some carry a full outfit of ready-made tools, while others obtain equally good results with a mere skeleton kit supplemented by odds and ends of specially improvised brushes, fashioned (after considerable experience) to produce distinctive effects.

'Rubbing-in' brushes

Employed for the rubbing-in or initial spreading of the graining colour. Two sets will be required, one for oil and one for water-colour work; each set consists of two ordinary flat paint brushes of 50 mm and 25 mm widths.

Mixing brush

As has already been stated, a half-worn 25 mm brush is ideal for the purpose.

Jamb duster

Some grainers carry two dusting brushes, reserving the part-worn brush for the preliminary streaking of certain straight-grain effects.

The flogger

A thin, long-haired brush which produces a suggestion of pores as seen in various hardwoods: the newly-applied oil or water stain is smartly stippled with the side of the brush, each panel being treated systematically from the base upwards.

Softeners

There are two varieties, one being of badger, the other being of hog's hair. The former is essential for the blending or softening of water graining, the latter being reserved for oil-colour work.

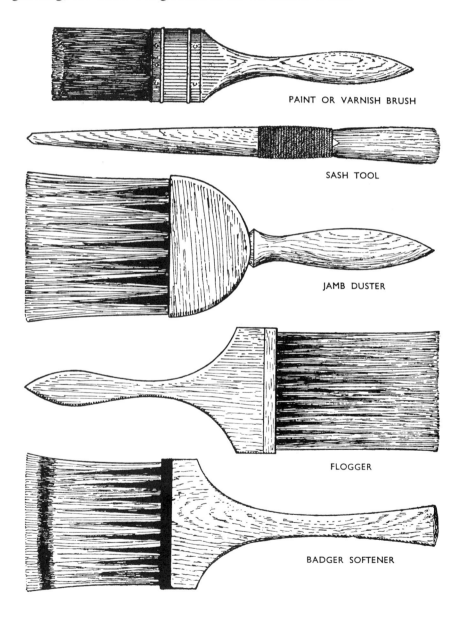

PAINT OR VARNISH BRUSH

SASH TOOL

JAMB DUSTER

FLOGGER

BADGER SOFTENER

Fig. 1. Graining brushes.

Fig. 2. Patent rubber graining appliance. This is dragged with a rocking motion through the wet colour.

Mottlers

Varying width and thickness give excellent reproductions of the highlights and shades which are a characteristic feature of natural wood. Mottling is usually executed in a water medium, but the beginner should explore its possibilities in oil-colour work, experimenting with the standard type and also with the stiffer home-made mottler prepared by cutting down an old paint brush.

Overgrainers

These thin, long-haired brushes are great time-savers and are equally efficient in oil or water media; the sizes commonly used are 40 mm and 60 mm in width. This type of brush produces parallel lines, the thickness being formed and regulated by passing the charged bristles through a coarse hair comb. The difficulty of maintaining an even tone may be minimised by using the brush at an angle of about 90° with the painted surface.

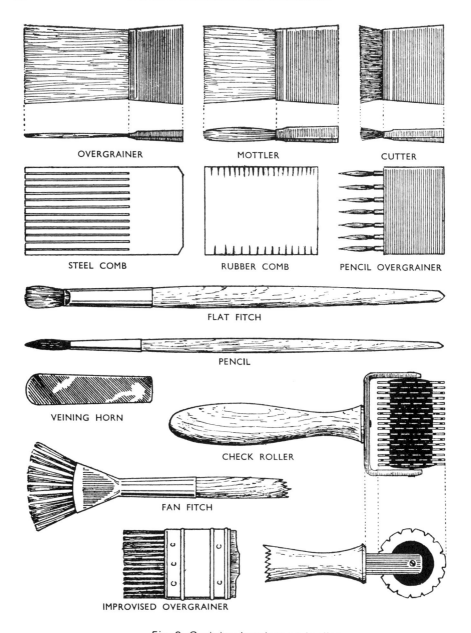

OVERGRAINER   MOTTLER   CUTTER

STEEL COMB   RUBBER COMB   PENCIL OVERGRAINER

FLAT FITCH

PENCIL

VEINING HORN   CHECK ROLLER

FAN FITCH

IMPROVISED OVERGRAINER

Fig. 3. Graining brushes and roller.

Pencil overgrainers

The best type is of sable hair with the pencils spaced to give a slight variation of widths between lines. This tool is ideal for the production of fine, widely spaced lines similar to those found in American walnut.

Camel-hair cutter or mottler

By means of this short-haired brush we can produce the fine, clean highlights which characterise feathered mahogany.

The fan fitch

Specially designed for the graining of pollard oak and other curly woods. It is of hog's hair and is used in an oil medium: considerable practice is necessary to acquire dexterity in its use. Many grainers prefer to make their own brushes or to adapt an old 40 mm paint brush into some semblance of a thin pencil overgrainer: such tools are capable of producing some extremely good results.

Maple dotters

Employed in the graining of bird's-eye maple. These can be obtained ready made or may be fashioned by cutting down a medium-sized camel-hair brush and then burning out a hollow centre with a red-hot skewer.

Fitches and sable pencils

Various sizes and shapes are necessary for adding the finer details of figuring or pattern.

Combs

Particularly valuable for the graining of oak, pitch-pine, and ash. The varieties in general use include several grades of steel combs and others which are cut by the grainer from rubber, leather, gutta-percha, celluloid, or cork. Each type has quite definite uses, which will be dealt with in the appropriate chapter.

Check rollers

Used instead of combs for suggesting the dark-coloured pores in oak. Some of these are equipped with a special clip-on type of brush which moistens the serrated discs during the printing operation.

Veining horn or thumb-piece

Used in conjunction with clean rag for wiping out the figure of oak, pitch pine, etc. It offers an efficient alternative to the grainer's thumb-nail, and is more hygienic and comfortable (see fig. 3).

Other useful items include a chamois leather and sponges for water-colour work, several goose feathers for walnut graining, an enamelled plate, and one or two large paint kettles.

Patent graining appliances

Various types have been devised from time to time, the most common being made of rubber which produce either a heartwood or straight grained effect found in woods such as pine and oak (see fig. 2).

Chapter Five

Combing, brush-graining and flogging

The first process is that of spreading or 'rubbing-in' the graining colour, but before this is attempted make sure that the ground colour is quite firm and free from defects. To apply graining colour upon a soft ground is inviting trouble, for the first strokes with a steel comb would tear up the paint beneath; similarly, any fat edges would fail to withstand the brush pressure.

It is equally impossible to obtain good results with rubber or leather combs if the under surface is ropy or dented. If the comb is to function properly its teeth must remain in firm contact with, and work smoothly upon, every part of the ground. It cannot remove stain from hollow or rough places. From this it is clear that good combing necessitates good preparation, and in the absence of the right conditions even the most elementary brush-graining would fail.

The application of oil graining colour

The consistency of the colour should be regulated until it can be easily applied with a 50 mm paint brush. This and any final adjustment to colour are ascertained by trial. If the graining involves the use of combs, try out their effect, for colour which appears to be of the right tone will often prove to be too dark when pulled into sharper and more compact lines by the rubber comb.

Steel combs are also of considerable assistance when making these preliminary tests, and it will be obvious, even to the inexperienced, that any crude or harsh effect is due entirely to the application of an excessive amount of graining colour, or to undue contrast between the graining colour and its ground. With these defects corrected, the combing will appear unobtrusive and woody, the colour will spread evenly, and the effect should be clean.

Part-worn brushes should be used for spreading the colour; these work more easily and enable the colour to be applied as sparingly as is consistent with evenness of tone. Incidentally, the large brush should be used whenever possible, reserving the 25 mm brush for window-frames and other narrow widths.

When using very pale colours it is sometimes advantageous to apply dabs of colour with the small brush and use the larger brush for spreading purposes. For general use, the large brush should not (unless a large area is involved) be overloaded with colour; a shallow dip into the graining colour, followed by light pressure against the side of the paint kettle (to remove surplus), will charge the brush with enough colour for a normal door panel and its moulding.

When 'rubbing-in' a door we follow the same order as in painting (see fig. 4). We commence with the mouldings of the top right-hand panel, then the actual panel is stained, and any surplus colour in angles or quirks in the mouldings is stippled with a clean dusting brush until the colour is evenly distributed. Any flogging, brush-graining, or combing is immediately completed before 'rubbing-in' the remaining panels.

The next job depends upon the setting speed of the stain. We shall find that parts of the stiles adjoining mouldings will be stained, but it should be possible to remove this dark edge with the brush as we continue to grain, first the muntins and then the cross-rails, finishing each item before passing on to the next. By this time the colour on the stiles may be too 'set' to permit clean 'working up' by mere brush pressure. At this stage we must use clean rag moistened in white spirit to clean the areas affected. This requires great care, particularly at the joints of rails and stiles, for these should be left perfectly square and true, otherwise the desired sharpness of definition will be reduced.

Finally, the door edge or edges and then the stiles are stained and finished. It will be fitting at this stage to point out a defect sometimes noticeable in grained work. Some grainers pay insufficient attention to the position of the junction of rails and stiles; instead of following the real joints, they adopt the illogical procedure of suggesting a joint where it happens to be most convenient, often with a very bad effect.

The more efficient grainer employs a bevelled-edged rectangular metal plate of about 200 mm by 75 mm, with which to mask or protect one end of a rail whilst staining the adjoining stile. This method is of considerable assistance in the production of clean, sharp joints.

Another detail of importance concerns the removal of door or other fittings prior to graining. Apart from their preservation in a clean condition, it is only by following this course that sharp continuous grain is obtainable.

Brush technique is much the same in graining as in painting, the colour being brushed firmly in all directions until the area is evenly coated; the brush is then used in the direction of the grain, and if an even colour results, the work may be stippled, etc.; if uneven, the surface is 'crossed' – i.e., brushed at right angles – and then 'laid off' lengthwise as before.

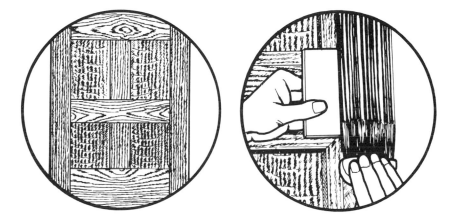

Fig. 4a. Oak grained door showing position of joints.

Fig. 4b. Use of a shield to obtain a clean, straight joint.

Combs and combing

Combing provides a simple method for the portrayal of coarse grain, particularly for oak and pitch pine. An examination of the plainer parts of these woods will reveal some interesting differences; pitch pine exhibits a certain boldness and continuity of line with fairly even balance between light and dark areas; oak is characterised by its vastly different ratio of lights and darks, the latter – which are actually broken-line effects produced by the pores – occupy something like one-fifth to one-tenth of the width of the lighter parts. It will therefore be evident that each requires its own particular combs and technique.

Rubber combs may be cut from waste pieces of linoleum, rubber or leather or other material of stout quality. These can be rectangular in form and of sizes varying from 50 mm by 75 mm upwards. Teeth are formed by cutting deep grooves of the required width on all four edges, but this must not be attempted until each edge has been accurately squared and straightened up: this latter point is important, as the comb will not wipe out cleanly unless the edge is maintained in a sharp and square condition (see fig. 3).

When forming the teeth, due regard must be paid to the effect desired.

In pine, for example, the notches and teeth are of equal width; in oak, the grooves are extremely narrow and the teeth are comparatively wide. In both cases it is desirable to make at least one comb with teeth on the several edges so graduated in width as to reproduce that natural coarse to fine variation in the graining (see fig. 5).

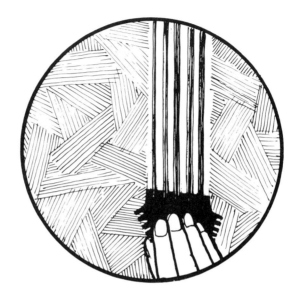

Fig. 5. Use of patent rubber comb with various sized teeth.

In pine, the combing is usually supplementary to the graining of heartwood; in oak, it may be similarly employed, or, alternately, it may be used as a background for quartered oak.

Combs should always be wiped clean after each stroke; this keeps the work clean, and helps to control the piling or ridging which would otherwise be too obvious in the finished work. In the case of oak, excessive piling of the colour produces an unpleasant effect just the opposite to that of real wood.

The broken character of oak grain can best be suggested by rubber combing, and then, whilst still wet, overcombing the work by a series of short, slanting strokes with a fine comb. Some grainers prefer to use a fine or medium steel comb with every third tooth broken out; others employ unbroken combs: both types should be tried out, varying the angles until one gains the power to reproduce coarse or fine pores at will (see plate 1).

Brush-graining

This is the term now applied to suggestions of 'grain' by the most elementary of all methods – i.e., by the simple expedient of drawing a clean dusting brush or 'drag' through the newly-applied graining colour. The result usually has little or no merit as interesting decoration except perhaps from the colour point of view, and even if the work is executed by a competent craftsman there is a feeling of mechanical repetition, and sometimes of harshness in the general effect.

Brush-graining can best be exploited by the grainer who normally makes use of plain graining either as a background for superimposed pattern or as a foil in support of figured panels. Harshness can be controlled by the judicious use of a badger or hog's-hair softener, and some variety in colour and grain will accrue from the use of brushes of various types. A painter's duster, for instance, will produce coarser and more contrasting grain than a half-worn paint brush or hog's-hair softener. Additional variety is obtainable by reversing the usual direction of the brush stroke, i.e., from a dragging to a forward scraping movement, the latter producing an effect rather similar to flogging (see plate 1).

These simple backgrounds are often used and broken by fine steel combing prior to the graining of quartered oak. On the other hand, the combing may be omitted and a suggestion of pores obtained by means of the check roller (see fig. 3).

Flogging and stippling

The pores form a distinctive and sometimes a prominent feature of the majority of hardwoods; it is therefore advisable to gain the requisite skill in methods whereby they may be reproduced. Here again is scope for experimental work on a scale wider than is generally realised and which must embrace a number of tools and materials.

Flogging provides a useful means of suggesting pores varying from coarse to medium according to the type of brush and graining colour employed. Coarse effects are obtained by flogging – i.e., vigorously stippling with the side of the brush upon the freshly applied graining colour. If flogging is delayed until the colour has become set, the pores will be smaller in size and quieter in effect. The maximum tone contrast is obtained as in brush-graining by manipulating a liberal coating of water graining colour (see fig. 6).

Finer grain is produced by brushes containing shorter, softer, or

Fig. 6. Producing pore marks by flogging. The arrow shows the direction the brush beats across the work.

more compact hair, and grainers should familiarise themselves with the effects obtainable by means of the badger softener, hog's-hair softener, and also by new and part-worn paint brushes (see plate 1).

Experimental work in various media is of fundamental importance as the experience gained by serious study of what one might term the preliminaries of the craft will contribute in no small measure towards ultimate success. This implies that time should not be wasted upon any haphazard approach to the subject, but that one should always work towards a definite objective – the reproduction of some detail of natural wood.

Stippling provides an alternative to flogging when finer pores are to be represented. The badger softener is probably the best tool for the purpose, with the hog's-hair softener and jamb duster following in order of merit. This light dabbing with the tip of the brush assists in the even distribution of the colour, and produces the slight texture effect which forms a groundwork suitable for mahogany or American walnut. An oil medium will provide softer and finer effects than those obtained in watercolour, but in either case the work may be lightly flicked with the badger softener if a quieter and more streaky result is desired.

Another method which is worthy of note consists of picking up the colour from a palette board and applying this by lightly stippling upon a dry or semi-dry surface. The tool employed is a badger-hair softener and, if used in a rather dry condition, exceedingly fine work is possible. In this instance we actually stencil the pores upon the painted surfaces and are better able to control their disposition, thus producing either close or widely spaced formations at will. This technique is particularly valuable when graining figured walnut.

A similar impression may be obtained by 'spattering' the work with colour which is brushed against and through the teeth of an ordinary comb. The effect will be more or less pronounced according to the consistency of the colour and the distance between the comb and the surface treated. It is usual to supplement this method by an immediate light dragging action with the badger softener in one direction to pull the specks of colour into elongated pore marks.

Overgraining

The value of a thin overgrainer for the production of simple straight grain is fully appreciated, but its possibilities in another direction – the suggestion of pores – are not generally recognised. It will be sufficient at this stage to point out that by overgraining in thin oil colour upon a dry surface, the lines thus formed may be broken by oblique strokes with a steel graining comb until a creditable representation is obtained. This technique is sometimes employed instead of the check roller for the production of coarse pores in oak, etc.

Chapter Six

The graining of ash, Hungarian ash, and light oak

Ash

Ground colour: Tint of BS 08.C.35.
Graining colour: Raw umber; raw sienna, and a touch of black.
Tools required: 'Rubbing-in' brush, combs of rubber and steel, clean cotton rag, veining horn, hog's-hair softener.

Ash is a light-toned wood of considerable decorative value. The grain is fine and compact and might well be mistaken for light oak; indeed, it requires a discriminating eye to recognise the smoothly rounded character and absence of rays which distinguish ash from the less serrated types of oak heartwood. Both are universally employed in the making of furniture, and also in wall panelling and other constructional work.

This wood is usually grained in oil colour, the figure being formed either by the removal of the colour by wiping out and combing, or by painting-in the pattern. To avoid repetition both methods will be described, the first being applied to our immediate project, the graining of ash; the second, to its more decorative relative, Hungarian ash.

Graining colour

This is prepared from the stock oil medium as previously specified, and is only slightly stained with umber and a touch of sienna. Cool silvery grey lacks the warm, cheerful tones obtained from the mixture given above, but its neutral quality is not without charm. A cool effect is achieved by greying or toning down the colour with a little black.

The colour should be applied evenly, and at the same time should be well brushed out; any shadiness or accumulation of colour in mouldings is removable by stippling or flogging with the softener. This done, the work is ready for figuring.

Wiping out the figure

For the wiping out of heart grain we fold a double thickness of clean rag over the veining horn, gather the loose ends, and hold firmly together with the other hand. This is a two-handed job and although one may at first feel extremely awkward when sketching the pattern with these unfamiliar implements, it will not be long before some measure of control is acquired. An expert grainer continues to wipe out with the right hand, while frequently giving a sharp pull with the left in order to maintain a clean working edge, all without cessation of work. Cleanliness and sharpness of definition are important factors which cannot be obtained except by the constant changing and tautness of the rag which covers the veining tool (see fig. 7).

Many grainers use the thumb-nail and are quite convinced as to its superiority in all types of wiped-out figuring. It can, however, become a painful operation after several days of continuous work, and in any case it is less hygienic than the method recommended above.

The handiest veining horns are those about 100 mm in length. These are already shaped; round at one end, fairly square at the other; and it is the square edge which gives the greater variety of shapes. In wiping out the hearts it is advisable to use the tool at the same angle as one would use a flat fitch, i.e., with the square end parallel to the direction of the grain. The long straight lines are wiped out with the angle nearest the direction followed, and by maintaining the firmest pressure at this point, we can, when forming the elliptical ends of curves, produce the required thickness of line and at the same time leave the outer edge sharp and the inner edge fairly soft (see fig. 8).

Fig. 7. Wiping out quartered oak figuring (silver grain) using a veining horn covered with rag.

Fig. 8. Wiping out oak heartwood figuring using a veining horn covered with rag.

Ash, in common with oak, displays more lights than darks, indeed, the dark areas are comparatively small. This feature must be consistently observed, particularly when straightening up the outer edges of heart grain, otherwise an unnatural variation – instead of graduation – will occur at the point of junction between lines wiped out and the auxiliary lines which are immediately executed by rubber combing (see plate 3 and fig. 15).

The next process, after the completion of wiping and combing, is that of softening. For this part of the work, we fold the rag to form a 12 mm-wide strip of five or six thicknesses. This in turn is folded to present a smooth face, 12 mm by 6 mm, and of reasonable firmness. With the pad so formed we stipple only the inside edges of the broad 'lights' – i.e., the curved portions – until the hard contrast between lights and darks is transformed into perfect graduation of tone. If properly done, the original lines will each present a hard outer edge and a soft inside edge.

Another method, which is especially useful for fine or close figuring, consists of softening with a dry brush, each stroke to be made towards the outer or sharper edges of the curves. A few light strokes will increase this sharpness of the one edge, and have a softening effect upon the other. Each light edge then appears to lift and overlap the adjoining dark edge.

The work is finally treated with the fine steel comb to produce serrulated lines as in oak. Any further strenghthening should now be completed with the addition of fine dark lines, which are painted in the darker parts of the curves to accentuate the lights. Stain used for 'rubbing-in' – especially if taken from the bottom of the paint kettle – is usually dark enough for this purpose, and only the lightest of softening is necessary.

Overgraining

Graining colours: As before, but ground in water.
Tools required: 'Rubbing-in' brush, sponge, mottler, fitch, and badger softener.
Medium: Stale beer, diluted with water.

Although overgraining is by no means an essential process in the graining of this particular wood, it does combine certain advantages which may be summarised as follows: it provides an opportunity for any minor correction or addition to form or colour; it enhances depth and translucency and thereby assists in countering any tendency towards paintiness; it enables work grained in water media to be fixed by overgraining in oil, thus dispensing with the usual strapping down of varnish.

Preparing the surface

It will be obvious that water colour will not 'take' very well upon a surface grained in oil, and that some preparation is called for. The best and safest method is to rub over the whole work with a damp sponge lightly powdered with fuller's earth or whiting to remove any surface grease.

Glaze colour

Should be relatively thin, its purpose being that of a glaze rather than a stain. In mixing, follow the procedure as for oil colour, i.e., adding colour to the medium, until the depth of tone is just sufficient to give a faint impression when mottled. Most beginners – and some craftsmen – make the glaze colour far too dark, and in consequence the finished work appears to be crude and muddy. As a general rule the glaze should be cooler and greyer than the colour beneath, and, being water colour, should be mixed in an enamelled or other rustless vessel.

Mottling

Apply the glaze colours with a clean 'rubbing-in' brush and reserve the mottler or cutter for its own special job of removing ribbon-like highlights or large areas interspersed with darker shades. If stronger tones are required, these can be applied with the fitch, but the whole work must be done expeditiously, to allow time for softening. Do most of the softening in a horizontal direction and complete with a few vertical strokes. Do not overdo the mottling, but seek inspiration from the natural wood, and it will soon be evident that light and shade are most pronounced in those parts of the grain which twist and curl away from the main direction (see fig. 9).

Complete any panels, one by one, wiping the mouldings clean with a damp sponge. Follow the same procedure with the mouldings, leaving stiles and rails clean. Finally, complete the muntins, cross-rails, door edges, and stiles; cleaning and squaring up the ends of each unit as the work proceeds.

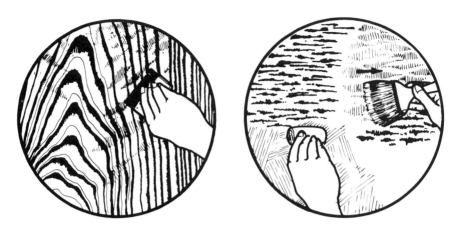

Fig. 9. Mottling using a squirrel hair cutter to remove the colour.

Fig. 10. Mottling using a wedge of rag to remove the colour. Softening is carried out in the direction of the mottles.

Note: Ash may also be grained in a water medium by the following methods:

(1) Prepare the surface with fuller's earth and a damp sponge.
(2) Paint the heart grain with a 6 mm one-stroke sable pencil, using the badger softener from the centre outwards.
(3) Put in the straight grain with a thin overgrainer, then comb to give a serrated appearance.
(4) Overgrain and mottle in thin oil-glaze colour.

Hungarian ash

Ground colour: Tint of BS 08.C.35.
Graining colour: Raw umber and raw sienna.
Tools required: 'Rubbing-in' brush, one-stroke pencil, No. 2 sable pencil, hog's hair softener, fine or medium grade broken steel comb, thin over-grainer, coarse hair comb, mottler, and clean rag.

This beautiful silky wood is slightly deeper and richer in colour than common ash, but its outstanding features are the highly decorative and curly nature of the grain and the striking display of highlights and shades. As might be expected, this wood is employed mainly for panelling, with less expensive and plainer material for rails and stiles (see plates 3 and 5).

To represent such intricate figuring by any wipe-out method would be too tedious a job, and the result would scarcely justify the time spent. Such fine grain is reproduced more speedily and efficiently by the opposite process – that of painting the pattern upon a suitably mottled ground.

Graining colour

Prepared as before, but with the addition of more raw sienna. This may be applied to the door panels, which are then mottled by horizontal strokes with a piece of rag folded over the back of the fingers, or, alternatively, with an old mottler (see fig. 10). Soften as in the overgraining of ash, and lightly flog. Mouldings and stiles may next be rubbed-in and grained in a plainer manner.

Graining

Carried out with a slightly deeper tone of the first colour. It is advantageous to allow the mottling to become dry, or almost dry, before continuing. In this condition there is no danger of the stain lifting or being otherwise marked when softening or combing.

The main figure is applied with a No. 2 sable pencil, keeping the lines fairly even in thickness and following an open formation. Introduce plenty of curl and smooth rounded shapes, observing the rhythmic disposition so conspicuous in the real wood. The work is then flogged lightly, and the supporting grain at each side is pencilled in. For wide panels it is advisable to use the sable pencil overgrainer, following closely and running parallel with the shapes already set. On completion, take a medium-grade steel comb and break up the figuring by a series of straight lines running the full length of the panel. Finally, put in any fine lines which may be thought necessary to accentuate the whorls.

Overgraining can be employed, but should not be necessary if the initial mottling is properly done; the work will already give an impression of adequate depth, a quality which is always improved by varnishing.

Oak, heartwood or sap

Ground colour, graining colour, and tools: As in ash.

Although many examples are similar to ash, there are certain characteristics which are only to be found in heart of oak. These differences must be grasped before they can be freely exploited by the grainer. The 'oak sap' – as it is frequently, and erroneously, called – shows considerably more ruggedness and variety of shape. The ends or outer edges of the concentric elliptical curves are sharply spiked, widely varied in thickness, and, as if in a final effort to break loose, the ends assume the most fantastic twists and turns, no two being exactly alike (see plate 4).

If we examine a 3 metre oak plank we should probably find group after group of roughly elliptical shapes, some large, others small, but all more or less in line. The spaces or junctions between these groups consist of plain and rather coarse grain extending for not less than 75 mm. There is no need to memorise the whorled shapes, as their reproduction presents no great problem. The main points for consideration concern the eccentric ends which terminate each succession of curves. These, or at least the ones selected as worthy of reproduction, should be observed in detail and recorded by carefully made sketches, which should be supplemented by others whenever opportunity allows. There is no better method of 'fixing' one's impressions, but note carefully whether the dark or the light parts are sketched; it makes all the difference between success and failure when the time comes to translate the idea into paint.

The heartwood is mainly employed in situations where it would normally be used by the wood-worker, i.e., for cross rails of doors or other work secondary in importance to the choicer quartered oak panelling. If the grained work is to appear natural, it should never give the impression of being 'centred'. This can so easily be avoided; indeed, where short lengths of timber are concerned it is better to omit the central curve and display the more interesting features of the wood.

A word of caution becomes necessary at this stage: be careful when graining the more intricate end shapes, as these, if overdeveloped, will upset the balance of the whole composition. It is not easy to lay down any hard-and-fast rule, but we would strongly urge all beginners to keep the extreme widths of such parts noticeably narrower than the points from which they are developed. By this means we are able to maintain the characteristic tapering formation which is common to most woods.

Method

The medium and the method for the graining of oak may be selected from the following:

(1) *Oil medium*, figure wiped out as in ash.

(2) *Oil medium*, figure painted as in ash.

(3) *Oil medium*, figure produced entirely by combing.

(4) *Oil medium*, figure sketched in crayon and blended.

(5) *Water medium*, figure painted, softened, and combed.

(6) *Spirit medium*, figure taken out with turpentine.

Of these, the first two are the methods usually adopted, and it is with these that we shall now concern ourselves.

In the first method, we follow the same procedure as for graining ash in oil colour. The work is rubbed in, lightly stippled or flogged, and the pattern wiped out with veining horn and rag. The coarse serrated end of each curve can best be suggested by a series of closely packed, zig-zag lines, the inside edge being generally smoothed, thickened, and cleaned up to produce an even colour, the outside edge being touched up in any parts lacking the required sharpness.

End shapes involve the use of broader lines, and, to facilitate clean wiping, it may be necessary to use the rag four-fold. In other parts of the work, it is advisable to graduate the thickness of line; this enables the heart grain to be straightened, and thus merge naturally into the rubber combing which flanks each side. Should the point of junction be noticeable, it may often be rectified by overcombing with a medium-grade broken steel comb (see plate 2).

Graduation of thickness is particularly necessary when working around any small spaces reserved for knots. In other woods the slight curl would create a subsidiary centre of interest, but in oak the grain levels out and is often running quite straight within 25 mm of the knot. The knot itself is painted in with a thin wash of the graining colours, slightly tinted with burnt sienna.

The next process will be that of softening, and although this is executed with clean rag as in ash, we shall have to contend with larger areas; a job which cannot be satisfactorily accomplished until the colour is partly set. Finally, the whole is slightly softened, the coarse grain is cross-combed, and the curved ends are broken or serrated with a 25 mm medium steel comb. In carrying out the latter movement, the comb is drawn along the centre in a series of short lines, each curving outwards; the general direction being somewhat like a rising column of water as it springs upwards and outwards from a fountain.

Overgraining and mottling

Executed in water colour as specified previously. The process is always advisable for heartwood, because it neutralises any tendency towards rawness, and the improvement effected does more than justify the little extra time entailed. Some grainers omit overgraining entirely, in which case the mottling is done upon a freshly grained surface by means of rag held four-fold over the back edge of a 75 mm steel comb. In print, this idea may appear crude; in practice, it can be exceedingly effective.

Check rolling

May be employed at the grainer's discretion when it is desired to accentuate the pores; otherwise it is unnecessary. The roller will behave equally well with an oil or a water medium. The colours may be selected from raw or burnt umber in oil, or Vandyke brown in water, according to the depth of tone required.

The roller is fed with colour from a wide mottler or brush which must rest lightly upon the upper edges of the serrated discs. The roller is moved forwards in the direction of the grain, leaving the pore-marks clearly defined. Too much pressure with the brush has the effect of removing colour and leaving the discs clean. If colour is too thin, the result will show an ugly series of blobs. Yet another case where success depends entirely upon practice (see plate 1).

In panelled work there will be a narrow strip at each end, which must be completed by a small sable writer. On completion, one final job must immediately be done – that is, the wiping of pore-marks from the light-coloured end shapes. This takes but a few minutes, using the rag and veining horn or, if rolled in water colour, using a damp wash-leather (see plate 3).

An alternative to check rolling has already been described (see 'Overgraining', Chapter Five).

The painting-in method

This will appeal most strongly to those capable of handling a one-stroke writing pencil; the method is rather quicker than wiping out, and, with practice, some excellent results are obtainable. Beginners are advised to avoid elaborate formations until sufficient skill is attained.
The process may be outlined as follows:

(1) Rub-in and streak with a coarse steel comb.
(2) Cross-comb with the medium broken comb.
(3) Paint in the figure, using a darker tone of the graining colour.

(4) Add the straight grain, using a thin overgrainer and coarse hair comb.

(5) Soften with a dry, soft brush to sharpen the curved ends of the figure.

(6) Break up the painted lines with the broken comb.

Complete the work by overgraining, rolling, etc., as desired.

Fig. 11. Use of a one stroke brush to paint in the heartgrain.

Chapter Seven

American walnut; teak; medium fumed oak

American walnut

Ground colour:　BS 06.C.37.
Graining colour:　Vandyke brown, burnt umber, and black.
Tools required:　Rubbing-in brushes; sable pencil, badger softener, mottler, pencil overgrainer.

In this chapter we are dealing with three varieties of timber, selected because of a certain similarity in colour and in the methods employed for their reproduction. Beyond these common factors there lies a vast difference, and our immediate problem is how best to render the essential characteristics by variance of technique alone. To summarise the more pronounced features, we should, by a brief survey of the real woods, gain the following impressions.

American walnut displays considerable depth and clarity; the grain is fine, distinct, retiring, and rather widely spaced; mottling is distinct and the pores are extremely fine.

Teak presents a rugged, fibrous appearance, the figuring being coarsely serrated, and, in consequence, much of the contrast and sharpness of definition is lost; the pores are long and coarse and there is little mottling (see plate 3).

Oak grain shows greater contrast than either of the foregoing woods, but, as has already been stressed, this quality can and should be controlled to give the work a general sense of repose.

The various species of American walnut exhibit greater differences of colour than of pattern. The grain is straight and smooth, variations being due to scale rather than to any inherent qualities. Some varieties are of a medium yellowish tone – gained by the addition of raw sienna – others, notably 'black walnut', are of a deep warm brown: all are liable to contain very slight traces of grey-green.

There are three distinct methods, all capable of giving a faithful representation of American walnut; each will be adequately explained, and, to follow the procedure adopted throughout this book, the process first described is in every case the one employed in graining the samples illustrated.

In the first method, which, incidentally, is that preferred by the writer, the work is executed in three stages, involving the use of water, oil, and water media respectively.

Stage one aims at the production of both mottling and fine pores. A thin glaze is prepared from equal parts of burnt umber and Vandyke brown, thinned with dilute vinegar or beer; prepare the surface, 'rub-in', and flog lightly with a clean, flat paint-brush until the marks obtained are crisp and fine. If any difficulty is experienced, try the alternative of stippling with the badger softener, then, whilst the colour is still wet, put in the mottle, and soften by careful stippling.

After an interval of half an hour or so, the work will be dry enough to withstand a coat of oil glaze without 'lifting' or 'working up'. The colour is prepared from burnt umber and should be little more than a glaze colour: apply this evenly and with very light brush strokes so as not to move the underlying colour; when partly set, we may proceed with the figuring.

The tools and materials required are: Badger softener, No. 2 sable pencil, pencil overgrainer, a little turpentine stained with umber, palette board, and graining colour darkened with burnt umber. Knots are usually left out of these straight-grained woods, but, should the grainer wish, these may be 'painted in' with a round fitch and some stain toned with burnt sienna.

When painting the figure it is advisable to use the brush in a rather dry condition, otherwise the colour may spread and become difficult to control. By dipping the pencil into the stain and then removing surplus colour upon the palette board, we are able to shape the pencil into a chisel edge, or point, as desired, and to maintain an even amount of colour in the brush. This will help to ensure greater unity in the finished graining.

The following are certain principles which affect the production of good design:

(1) Observe the smooth gradual tapering of the heart grain as it extends upwards from the base and note the shallow elliptical curve sweeping inwards as it ascends.

(2) Avoid the repetition of hollow shapes – i.e., lines directly opposed to those mentioned above. There will, of course, be an occasional

hollow which has crept inadvertently into one of the tapering curves; such lines can be allowed to remain, but the shape should be corrected when painting the next line.

(3) Try to vary the spaces between the curved ends: some may be closely grouped, others wide apart, but on no account must there be any suggestion of equal measured distances in the finished work.

(4) Ends should be of varied shapes, some angular, others curved; avoid acutely pointed formations (see fig. 12).

Fig. 12. Producing the gradual tapering heartgrain using a fan fitch. This shape is the basic form of most heartwoods.

It is now assumed that the central figure has been satisfactorily completed; the sides should be grained either by the pencil overgrainer or, if the panel is narrow, by the small writing pencil. Then dip the pencil into the glaze colour, use the palette board to remove surplus material, and then run a fine line in some of the spaces between the end shapes. Soften lightly with the 'badger', always brushing towards the ends until the combined action of turpentine and softener gives the required half-tones; any harshness may be toned down by lightly flogging or stippling (see plate 10).

When dry, the work is glazed with Vandyke brown and a touch of black; some of the water colour left over from stage one can be thinned down and the necessary adjustment made to colour. The underlying mottle is now picked out again and any further lights or shades may be

Plate 5

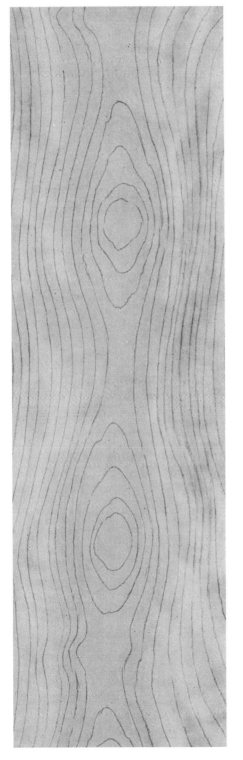

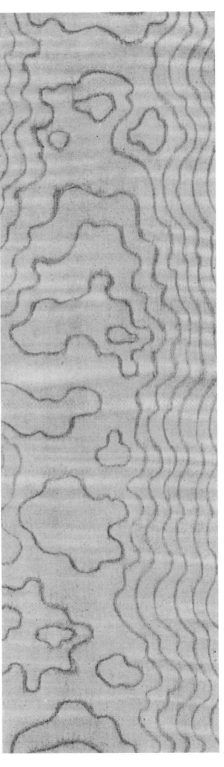

Satinwood

Hungarian ash

Plate 6

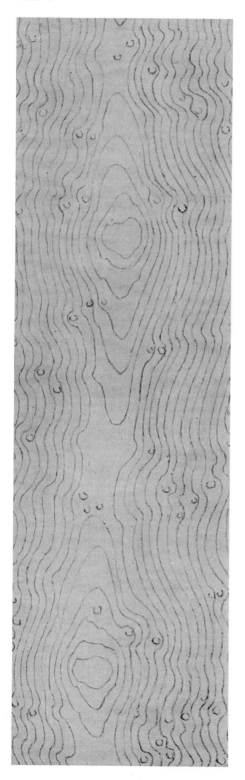

Bird's-eye maple

Pine

Plate 7

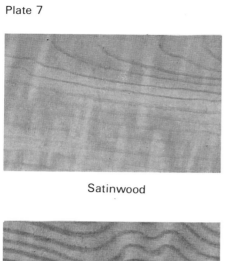

Satinwood

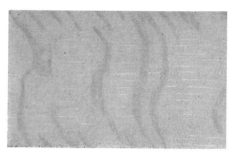

Limed oak

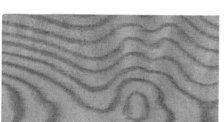

Pitch pine

Fumed oak

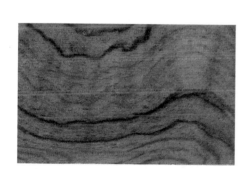

Walnut

Burr walnut

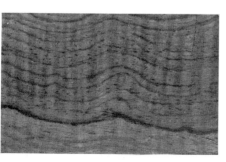

Italian walnut

Antique oak

Plate 8

Root of oak   Rosewood

added to the plainer parts of the grain. Complete by softening until the mottle is just discernible.

Water colour method

This is quite sound and expeditious, but hardly suitable for the beginner who has not yet mastered the technique. The first stage is carried out as before, i.e., by 'rubbing-in', mottling, and flogging, and, when dry, the whole is grained in the same medium, using tools and pigments as before.

We commence by sketching in the central heart grain with a pencil overgrainer charged very sparingly with an extremely pale tone of the colour. This is softened in a upward direction to form the half-tones which are a subsidiary feature of the grain. To obtain the requisite quietness of effect we must not oversoften and thereby produce hard, contrasting edges: it may even be found that water alone is sufficient for the more shadowy parts, but this must be left to the judgment of the individual.

The main figure is painted in with the sable pencil and a darker tone of colour. It is merely a matter of following and strengthening some of the existing lines, softening but not spreading the colour as the work proceeds. A common difficulty is that of preserving an even colour, and in this connection we suggest frequent stirring of the colours and the use of an odd plate or palette board as previously indicated.

Straight grain is painted in with a thin overgrainer or sable pencil overgrainer (see fig. 13), the latter being more easily controlled and giving the thin, widely spaced lines consistent with the centre figure. The thin overgrainer can, of course, be divided by a coarse comb, but to secure the best results it should, for this graining, have been worn to an exceedingly thin condition.

Overgraining is left entirely to the grainer's discretion. It should, however, be remembered that in addition to its primary purpose, an oil glaze serves as a fixative for the water graining beneath.

Graining in oil

This is a useful one-process method closely related to that employed for Hungarian ash. In this instance we can use the same tools, and with a slight variation of colours we produce American walnut. Such cases are

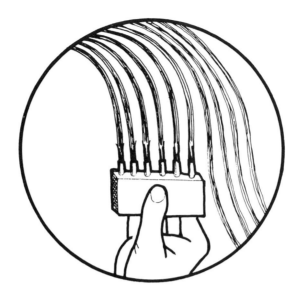

Fig. 13. Use of pencil overgrainer to produce widely spaced straight grain.

common in graining and marbling. Having mastered about half a dozen main types, we should experience no difficulty in the graining of others. At this stage we have only to see a sample of any wood and we are able to recognise the colour and technique necessary for its reproduction: fortunately , we are still able to appreciate its full beauty and – it is to be hoped – with a wider vision than before.

In the graining of walnut we require oil colours only, the work being 'rubbed-in' rather sparingly and then mottled and softened. When set, the delicate half-tones of the heartwood are put in with paint or crayon and lightly softened and flogged. The strong centre grain and plain supporting lines are added and the work is again softened and flogged. Finally, the mottle is accentuated, and if the flogging has failed to give an adequate suggestion of pores, we apply a little dark colour, stippling it upon the surface with the tip of the badger softener: the latter operation should be executed with the brush almost dry.

Teak

Ground colour: BS 08.C.35.
Graining colour: Burnt umber and burnt sienna.
Tools required: 'Rubbing-in' brushes, flogger, flat fitch, sable pencil, jamb duster, medium-grade broken steel comb, pencil overgrainer.

To gain a good representation of the fibrous character which is the predominant feature of teak, we cannot do better than to rub the work in with a water stain containing burnt umber and a little burnt sienna; this is to be heavily flogged until of coarse texture. When dry 'rub-in' with an oil glaze of similar colour, await partial setting, and then, with an old flat fitch, paint the heart grain in burnt umber thinned until only one or two shades deeper than the ground colour.

There should be nothing suggestive of laboured brushwork; the figure will have more character if executed boldly and sketchily with the brush held at about 45⁰ to the surface. Thin lines are obtained from the edge of the brush as it moves forward to produce the broad serrated curve; then, by keeping the fitch at the same angle, the thickness is automatically reduced as the brush edge comes into play on the downward stroke (see fig. 11).

The plain work on either side may be painted in with the fitch, or, if the area is wide, with the hog's-hair pencil overgrainer. For softening the work we require a clean jamb duster, preferably part-worn; then, with a series of light, dragging strokes, we blend and break up the graining. Further serration may be introduced by steel combing, taking care not to lift the water stain beneath. Finally, any faint mottle which may be thought necessary can be put in by light pressure with the padded edge of the steel comb.

Alternative methods

Include graining wholly in oil or water colour as convenient. In either case, the work is 'rubbed-in', flogged, grained, and roughly softened and flogged, all at the one operation. Tools and materials as indicated previously. The main recommendation of these short-cut methods is obviously that of speed, and whilst we must recognise the importance of the time factor, we must also realise that quality of workmanship is of first importance; speed being the natural outcome of mature experience.

Medium oak heartwood

Ground colour: BS 08.C.35.
Graining colour: Burnt umber, raw sienna, and a touch of black.
Tools required: 'Rubbing-in' brushes, jamb duster, one-stroke sable pencil, thin overgrainer, combs, rag, badger softener, mottler, check roller.

Medium oak exhibits much the same variations of colour as American walnut: some examples are cool; others incline to a warm brown and may be matched with burnt umber alone. For cool tones we can neutralise the umber by adding a little grey-green produced from raw sienna and Prussian blue, but this must not be carried too far (see plate 4 and fig. 15).

In graining, this wood may be used with good effect upon rails and stiles, its quiet colour serving as a useful foil to brighter panels of quartered oak, pollard oak, etc. For framing of this type it is preferable in every respect to the more assertive wiped-out heart which competes too strongly with the main centres of interest – the panels.

For the graining of medium oak we employ a method similar to that specified for teak. The stain is rubbed in evenly and rather sparingly, then flogged or stippled. Follow this by straight graining with a coarse steel comb and afterwards overcombing with the medium-grade broken comb.

Clean rag is in constant demand whenever combing operations are in progress; the two seconds spent in wiping the teeth after each stroke have an important bearing on cleanliness and sharpness of finish, and in any case, the comb is intended for the removal, not the application, of stain.

Another and equally relevant factor concerns the interval of time between 'rubbing-in' and combing. There must be at least a few minutes allowed for the initial setting of the graining colour, otherwise sharpness will be lost. In the present instance we have considerably more latitude than in rubber combing, and it is better to err on the safe side and not follow up the work too quickly; we can usually continue to 'rub-in' other sections of the work and thus avoid waste of time.

The combing is allowed to set before continuing to grain as in teak. We must, however, keep the lines finer and the grain more compact and employ the thin overgrainer for the straight grain. Follow by blending lightly with the dusting brush or other softener until the curved ends develop a sharp outer edge; at this stage we may, if desired, break up the work by several strokes with the medium split comb.

When dry, the work is prepared, overgrained in water colour, and completed by check rolling.

Alternative methods

(1) Execute the work as above, substituting crayon for the actual figuring.
(2) 'Rub-in' with water stain, and stipple. Grain upon the dry surface in oil colour; overgrain in water.
(3) Stipple, grain, and roll in water colours; overgrain and mottle in oil.

Each process should be fully explored, using graining colour of varying consistency; the experience will be of considerable value when, at some future date, one is required to match existing work.

Chapter Eight

Quartered oak: light figure, dark figure, and spirit grained

Ground colour: BS 08.C.35
Graining colour: Raw umber, raw sienna, black.
Tools required: As for ash.

It is generally acknowledged that quartered oak is the most difficult wood to grain. The grainer, in his search for samples, may find a bewildering range of nondescript pieces, quite useless as copy and worse than useless if they give a false or exaggerated first impression. The beginner *must* start right by working from good examples of English or other equally compact varieties of oak, avoiding the massive American oak, which rarely gives an adequate picture of the underlying shapes which govern design.

At this stage it is well worthwhile to discover some relationship between the familiar heart grain and the more complex figure presented by quartered oak. Both are obtainable from the same log of timber according to the direction of the saw cut (refer to fig. 14(a)).

All trees share certain common features; they are governed by the same elementary principles of growth which involve the formation of annular rings, the whole structure being a mass of minute longitudinal cells of varied thickness. Cells formed by the spring growth are comparatively large; those added in the autumn are small and compact. It is this latter cellular mass which, after years of compression and chemical change, produces the apparently solid timber.

The heart grain may result from either of two conditions. In trees of the fir type, which are quite straight, it is possible to saw down the whole length, and only cut through two or three of the more central rings. This would produce very little heart grain and would be of the most elongated pattern. If, on the other hand, the cut were made in a diagonal direction, the elliptical ends would be more compact.

In other trees growth is more irregular: prevailing winds affect straightness of growth, and in less sheltered positions cause twisted and

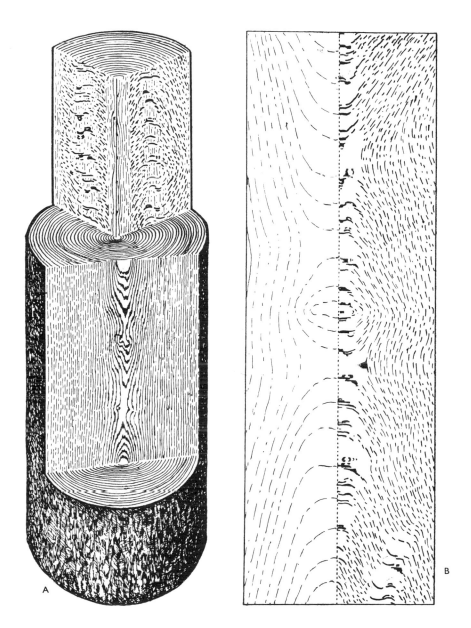

Fig. 14. (a) Oak, showing heartwood and silver grain; (b) the pattern basis of quartered oak.

curved formations. Branches and side-shoots also contribute in no small degree to the ultimate run of the grain. The saw cuts and re-cuts through the annular rings, and the more wavy the log, the greater and more compact will be the series of elliptical shapes produced.

In certain woods there are other groups of connecting cells which extend from the pith, or centre, through the body of the tree to the outer sapwood. In oak these rays are highly developed, extremely numerous, and clearly defined. Only the paper-like end sections are discernible when these are cut across to show the heart grain, but when a tree is sawn across the centre, or cut quarterly, or radially, in the direction of the rays, the silver grain is at once strikingly evident.

Upon examination, one may trace a well-defined sense of direction which appears to control the flow of these beautiful and interesting shapes. The lines, though considerably broken, will usually follow the general pattern presented by a rather squat example of heartwood. Changes of direction are not nearly so frequent, but are still clearly recognisable.

It is important that students, when examining a piece of oak, should look at and beyond the figuring, and endeavour to trace the less obvious laws governing pattern. In this way one's inventive faculty is provided with the most useful material.

The graining of oak (light figure)

Of the several methods used for the reproduction of this wood the oil-colour process is by far the most popular with present-day grainers. Work carried out in this medium is under complete control from start to finish, and there is no possibility of accidental effect creeping in and spoiling the appearance.

To the craftsman who knows the run of the grain and has learned to control the veining horn, the following directions should help in the production of a creditable job. In the event of failure the whole can be washed off with a clean rag and white spirit, and the work re-commenced – a job not possible in water colour.

And now one word about pattern. If we refer to fig. 14 it will be noticed that the lines indicating the silver grain – popularly described as the clashes or dapples – are not just segments of one continuous line, each following the next in single file. There is, of course, a common sense of direction, but the ends of the dapples will be found to overlap like so many isolated roof tiles, and it is this feature which must be continually observed even though the lines curve and change direction. Notice also the steep pitch of the slope, a positive advantage when twisting and curling the run or flow of the grain.

The main dapples – *i.e.*, those forming a right angle with the grain – are fairly angular and several times thicker than the fine work. These are characterised by the number of fine, horizontal breaks which elaborate the general appearance of what would otherwise be simple shapes. Long vertical lines are inclined to be too assertive and are better left out or, at least, broken into several shorter lines.

Spacing should be constantly varied by alternating open patches with more closely packed areas. There is little of interest in, and certainly something very unnatural about, any repetition of similar spaces.

The connection between large and small dapples should be made sufficiently obvious. Each forms a part of the one design or pattern and must take its place in the rhythmic flow which clearly establishes relationship.

The dotted line running down the centre of fig. 14(b) shows how the figuring may be simplified by division, one half being suitable for the right-hand, the other for a left-hand, door panel. Further simplification is advisable (though not absolutely necessary) during the early stages, and may be achieved by selecting any part of the pattern to form a basic shape which will decide the direction of flow for one whole panel. There need be no changes of direction, but only of spacing and thickness of line.

Even for an extra broad panel which calls for the full figuring, it is better to show only one or two changes of flow. The diagram is not intended as a model from which to copy, but for the more practical purpose of illustrating the more important changes to be seen in oak; each formation would be all the better if extended for 300 to 600 mm before changing direction.

Ground colour

This has been given as 08.C.35, yet it will be clearly understood that this must not be regarded as a hard-and-fast rule. There are many tints to be found in new or light oak, some a little darker than cream, others greyish, and some showing a slight tendency towards green. It is sufficient to note that the desired quality of tone should be present in the ground colour as well as in the graining colour.

Graining colour

For general purposes consists of raw umber with only a touch of sienna. For the more yellow tones add more sienna; for a neutral tone try umber alone; and for grey, see what can be produced from umber and black or sienna and black. The results are surprisingly varied and the experience provides useful data for future use.

Figuring

This should be carried out as soon as possible after 'rubbing-in' and combing. It is particularly important to secure a pale and quiet background for the silver grain. In fact, the finished work must show only a quiet tone contrast, which means that the graining colour must be extra pale and the rubber combing well broken up by cross combing. A contrasting background is not only contrary to nature, but will show up any poor figuring in a most conspicuous manner (see plates 2 and 3).

The dappling is wiped out with veining horn, or thumb-nail and clean rag, as in the graining of ash. The rag selected should be clean, soft, free from lint, and sufficiently absorbent. This need only be of single thickness when used to scoop out the silver grain.

It is better to commence at the top of a panel and work downwards, wiping out only the main dapples, and arranging these according to a preconceived plan or design. The whole of the figuring can be executed with the left-hand corner of a chisel-edged veining horn, held sideways, as if using a flat fitch. By this means we can vary the shape and thickness at will; a thin line formed by the corner of this thumb-piece will increase in width as we bring the flat chisel edge into play and will diminish just as easily when the left-hand angle is again brought into use. It is advisable to practise this stroke, keeping the chisel edge in a vertical position throughout, and using a slight rocking movement to bring the edge more or less into contact with the wet stain (see fig. 7).

Clean wiping is important, the rag being frequently moved to bring dry parts into use, otherwise it will be found that when saturated with colour, the rag will behave like a rubber comb, leaving prominent dark edges.

When wiping out the small dapples we employ the same corner of the thumb-piece, varying the angle as convenient, but always using the edge of the tool. The marks may be of different lengths, some – particularly where the grain curls – being little more than dots, but all must conform with the run or flow of the pattern. Throughout the process the rag is firmly held, the parts not in use being gathered up in the left hand so as not to drag upon the work.

Softening

This is carried out in three movements; the first, which concerns only the large dapples, consists of softening each in turn by lightly brushing along its whole length with a soft-haired fitch. The second stage is that of mottling to obtain a faint impression suggestive of secondary formations lying just below the surface. These are most evident in areas between the main dapples. Mottling is carried out by lightly stippling with the edge of a pad, obtained by folding about eight layers of rag. Stage three is a general one-way softening with a dry brush: by lightly brushing in a downward direction, a faintly discernible shadow can be formed along the bottom edge of each dapple.

Overgraining

May be carried out upon the dry, or almost dry, graining colour. The latter course has the advantage of saving time, but requires care and judgement, lest the undergraining be damaged. The work, if executed on the same day as the figuring, is done in oil colour; a little graining colour, thinned with turpentine, is applied with the thin overgrainer to form a series of faint lines running parallel with the coarse combing underneath. These lines are carefully broken up by cross-combing as in Hungarian ash; and, finally, the large dapples are wiped clean.

Oak: alternative method

(1) The graining colour is applied more liberally, and is then flogged or brush-grained.
(2) Wipe out the figuring and soften as before.
(3) Glaze and mottle in water colour.
(4) Add coarse pores by check roller and water colour, leaving large dapples clean.

Oak can also be grained in water, although the process is but rarely used. The general procedure is as follows:

'Rub-in' and comb or flog; wipe out the figuring, using a damp chamois leather over the thumb-piece. Unless one has attained the necessary proficiency and speed, it is better to delay the figuring until the initial colour is dry. The work is overgrained in oil colour, using a 6 mm one-stroke pencil.

Graining in oil is sometimes executed upon a background of

unpainted wood which has been rendered non-absorbent by several coatings of white knotting. Success depends very largely upon the quality of the timber, freedom from knots, etc. An interesting variation is secured by painting the figure in white knotting upon the untreated wood; afterwards staining in oil colour. After a short while remove surplus colour with a pad of clean rag; the dapples are left clean and sharp upon a tinted ground. Overgraining, supplemented, if desired, by check rolling, gives a satisfactory finish.

Quartered oak (dark figure)

Under certain conditions, which may be due either to age or to chemical action – such as the 'fuming' of oak – the figuring is considerably darkened. In some cases the contrast of tone is conspicuous, in others only slight; much depends upon the angle from which the wood is viewed. At times, the figure appears to be quite pale when examined from a certain position; then the tone changes with the view-point, and, when the lights are no longer reflective, the figure appears dark.

To represent such a combination of light- and dark-tone values would require material of almost magical properties; our ordinary palette imposes very definite limitations, the net result being that we must elect to represent only one of the several aspects presented.

In dark figuring we gain a perfectly legitimate variation of quartered oak, and, as this darkening is not limited to quartering, we might on some occasions combine this with fumed heartwood. The following methods may be considered advisable:

(1) Paint the dark dapples upon a surface previously 'rubbed in' and combed (in oil colour) and allowed to dry. The whole may be executed with a 6 mm one-stroke pencil, charged very sparingly with thin translucent colour. When completed, the whole is lightly 'badgered' in a downward direction, and, when dry, the work is over-grained in water and suitably mottled, etc. (Refer to Plate 7.)

Some variations worthy of trial are:

(2) Painting upon the semi-dry rubber combing, and thus speeding up the initial stages of the work. This method also permits some rag softening, but the 'badgering' calls for great care: if overdone, parts of the combing may be lifted.

(3) Dapples may be painted in oil directly upon the dry ground colour. When dry, the whole is glazed, mottled, and check rolled in water colour.

(4) This is similar to the first method, which requires a dry combed background. In this instance we paint the figure in watercolour and overgrain in an oil medium.

(5) Another variation that appears to have emanated from conflicting ideals attempts the simultaneous display of lights and darks. The effect, when properly controlled, is subtly quiet and at the same time interesting.

The work is first grained to represent quartered oak (light figure), the small figuring being softened more than is usual. Stage two consists of the immediate application of an extremely thin wash of burnt sienna upon the central portion of each large dapple. Softening is carried out with a soft fitch used in the direction of the figure. The whole is completed by overgraining, etc., in water.

Quartered oak (spirit grained)

Graining colour

For this work, which is certainly unusual, we must take a composition which will dry quickly and yet be capable of instant and complete removal when graining. Furthermore, it must be 'fixed' just sufficiently to withstand varnishing.

We begin by mixing some gilders or other fine whiting in turpentine until of a smooth consistency; tint by the addition of oil stainer and thin with more turpentine until of the right fluidity for easy brushing. The binder or fixative is japanner's gold-size, and the amount necessary must be ascertained by trial. Add only a little at a time, testing upon some small area of the work, until it is found to be reasonably well fixed. Strain the colour and use immediately, or store in an air-tight container for use the following day.

Graining

The first stage embraces the usual 'rubbing-in' and combing. It will, however, be necessary to work quickly because of the extraordinary drying properties of the colour. This job calls for first-time methods; as each panel is 'rubbed-in', it must immediately be combed, and there is no time for any intermediate stippling or flogging. The whiting is inclined to settle out, leaving the colour pale and washy, a point to be rectified by frequent stirring.

Dappling is executed with equal rapidity by the simple process of painting the shapes in turpentine or white spirit, and immediately removing the softened colour by a downward sweep with clean, soft rag. The process soon becomes a synchronised movement of painting with the right and wiping with the left hand.

The figuring is extremely clean, yet this, as well as the sharpness, is not uncontrollable. It will be found that the greater the pressure, the cleaner the colour, and that sharpness can be modified by two-way wiping, the downward sweep giving a soft shadow along the bottom edge of the dapple, the upward stroke, if carried out quickly, being sufficient to remove harshness in the upper edges. Overgraining in water colour completes the work.

A similar process of painting the dapples with white spirit and softening in one direction with a badger softener can be employed on wet oil graining colour. It is especially suitable for imitating the very fine dapples found towards the edges of planks of quartered oak.

Chapter Nine

Limed oak and weathered oak; decorative treatments

There are many occasions when, for one reason or another, something out of the ordinary is desirable. It may be that some question of colour is involved, or the mere seeking after novelty, but whatever the circumstances, if the problem concerns the treatment of woodwork, the grainer should, as a result of his specialised training, be in a position to offer a satisfactory solution.

Traditional methods and colourings will remain as long as wood continues to grow, yet we must recognise the charm of the more neutral colourings as presented in the various forms of limed and weathered oak, pickled pine, and others popularised by architects, furnishers, shop-fitters, and others, who create and stimulate a demand for brighter and more cheerful fitments.

Graphers are not prone to be impressed with the idea of novelty for its own sake, but in the midst of ever-changing conditions, it might be as well to relax certain rules, with the idea of becoming more enterprising.

Limed oak

Ground colour: BS 08.B.17.
Graining colour: White, raw umber, and black.
Tools required: As for quartered oak.

In the graining of this wood we reverse the procedure hitherto followed. Lime, whether applied accidentally or deliberately, remains in the pores of the timber, and also influences the general colouring according to the extent or period of saturation, the state of the lime, and the age of the oak, the result being a universal darkening of the figuring and lightening of open pores.

We must draw a sharp line between the genuine limed oak and limed oak effects obtained by the use of oil-bound distemper. The latter has no darkening effect on the figure, but rather the reverse. It imparts, particularly in the case of new oak, a soft, bleached appearance. Perhaps it is not surprising, in view of man's inherent desire for colour, to find articles of furniture treated, not with lime, but with pale greens, blues, etc., often of a soft metallic quality. Here is another promising field for experimental work by the grainer.

Graining colour

This should be used as thin as is consistent with the slight colouring required; by this means, we secure the maximum degree of translucency and avoid a hard, painty finish. The work is 'rubbed-in' and lightly flogged and then wiped out to produce the quartering or heartwood. In the latter case rubber combs are employed to obtain the straight side-grain and steel combs for serrating the elliptical ends.

Softening

Using the cloth pad, fitch, and softener, exactly as specified for oak (wiped out). The work is then allowed to dry.

Overgraining

May be limited to the simple process of check rolling with a pale tint of the graining colour. By keeping the feed brush and roller in a fairly dry condition it should be possible to avoid any exaggeration of these poremarks. The final operation consists of wiping all large dapples and ends of heart shapes; these are not usually broken by large pores (see plate 7).

Weathered oak

Ground colour: BS 00.A.01.
Graining colour: Black, burnt umber, ultramarine blue.
Tools required: As for quartered oak.

Weathering has a pronounced effect upon the natural colour of unprotected woodwork. The bleaching action of sunlight, together with periodical saturation with rain, tends to produce a very definite greyness and considerable reduction in contrast. Oak is no exception to the rule, but while colour is strongly affected, contrast is partly maintained. The lights, being hard and smooth, retain their reflective powers, and the pores, especially when coarse, collect various deposits which assist definition.

The colour is not easy to match, some examples being cool and silvery, while others are neutral grey of a slightly warm tone. It will be found that burnt umber and black will supply any desired warmth, and the addition of blue is very effective in the production of cool tones.

Graining is executed by the wiping-out method and completed precisely as in limed oak (see plate 4).

Decorative treatments

The foregoing text has explained some of the techniques which can be employed to imitate the colours and patterns of real woods. Once these techniques have been practiced and mastered, experiments with colours not normally associated with natural wood could be considered. By substituting the soft earth colours of natural woods for stronger colours such as bright yellows, blues and greens for both ground and graining colours, a range of interesting vibrant effects can be obtained.

The use of these colours in conjunction with flogging, mottling and other graining techniques could be used not as a method of imitation but purely as interesting broken colour effects to enhance an otherwise plain surface.

Chapter Ten

Pitch pine; pollard oak; root of oak

Pitch pine

Ground colour: BS 06.C.33.
Graining colour: Raw sienna; burnt sienna; burnt umber.
Tools required: 'Rubbing-in' brushes, one-stroke pencil, pencil over-
   grainer, badger softener, mottler, and fitch.

The colour of this wood, irrespective of its age, can be matched from the pigments given; raw sienna is the main ingredient of the graining colour, and burnt sienna toned with burnt umber will supply all that is needed for graining. In the final stages we shall also require raw and burnt sienna in water.

Graining

Best executed in oil colour, upon a surface which has been 'rubbed-in' sparingly and allowed to set. There is no preliminary brush-graining or stippling; in fact, all brush-marks should be obliterated by softening until the effect is plain and translucent.

The main figure of the heartwood is painted in with a semi-translucent colour, containing equal parts of the colours given above. This provides just the right amount of contrast with the raw sienna ground. A palette board and one-stroke pencil are particularly useful at this stage, for the colour must be applied evenly and thinly, with lines thin and closely spaced at the sides, and becoming broad and extended at the ends (see plate 7 and fig. 15).

In the plainer examples of heartwood, there is much repetition of elongated pattern similar to that of deal, but the choicer pieces are fairly compact, and full of smooth, rounded curves, quite as intricate as in Hungarian ash. The plainer side-grain is painted in with a sable pencil overgrainer – or thin overgrainer – carefully following the twists and curves set by the central pattern but gradually reducing these as the distance from the heart increases.

OAK (HEARTWOOD)   ASH   PITCH PINE

Fig. 15. Sketch of heartwoods.

Softening

This is executed with the badger-hair softener, the object being to brush out the darks so as to produce hard outer edges and soft inner edges, as in the case of fumed oak. It should be noticed that this effect is exactly opposite to that required when graining oak by the wipe-out method. For oak, we soften by a series of inward strokes; for pitch pine, in the reverse direction. Failure to observe this point has often ruined the appearance of what should have been good work. The plainer side-grain presents no problem, as this requires only the lightest up-and-down softening in the general direction of the grain.

Overgraining

This is carried out in a water medium containing equal parts of raw and burnt sienna. Only the thinnest glaze is necessary, the object being to introduce light and shade by mottling and painting in with the fitch. In pitch pine the lights constitute a prominent feature of the grain; they are frequently large and wedge-shaped, like cones of light pointing towards the central figure. As in other woods they are placed in the curly part of the figuring and carefully softened by the 'badger'.

Wiping-out method

This may also be used in the graining of this wood. The procedure is the same as in ash, omitting only the fine steel combing. In pitch pine, rag stippling is employed to soften the *outer* edges of the lights, brush softening being unnecessary. Overgrain in water, as described above.

Graining in water

Occasionally employed, but is not so easily controlled as oil colour; furthermore, it lacks both strength of colour and smoothness of form.

Pollard oak

Ground colour: BS 08.C.35.
Graining colour: Burnt umber; raw sienna; burnt sienna.
Tools required: 'Rubbing-in' brushes, pencil, fitch, steel and rubber
   combs, rag, veining horn, pencil overgrainer, hog's-hair overgrainer.

The eccentric and rather curly figuring of this wood is the direct outcome of man's deliberate interference with the natural growth of the tree. When the trunk has attained its full height and its growth is well established, the main branches are lopped off; this induces the formation of numerous side-shoots, each group forming small knots which cause considerable curling and twisting of the grain and some local darkening of colour. Thus it is from the upper part of the tree that this variety of oak is obtained (see plates 1 and 4).

It is usual, in graining, to reserve this type for the panels of doors, etc., where it forms an effective contrast to rails and stiles of plain oak. In representing this wood, we incorporate a certain amount of heartwood, quartering, combing, painting in, and running out with turpentine as in spirit graining, obviously a job requiring close study of more than one example of the real wood.

Graining colour

Prepared as for quartered oak, but should be a shade deeper in tone. This is 'rubbed-in' and manipulated to produce a coarse impression of grain, twisting around the groups of knots, and then joining up to flow smoothly on until other knots are encountered. For this stage of the work a crumpled piece of hessian or, better still, a worn-out small brush, will produce the contrast and accidental quality desired.

Graining

The knots may now be completed by the popular method of adding more stain to the areas concerned, and, with the rag folded to produce a sharp corner, the light parts surrounding the knots are lightly wiped out. It is particularly important to observe contrast of size and shape; there must be no exact repetition or the work will appear mechanical. We can, however, introduce variety by the partial outlining in darker colour of some and the sketching of fine, paler lines around other knots, as seen in the natural oak.

It will now be necessary to put in the combing before the graining colour has had time to set. With the rubber comb we add the straighter grain, afterwards cross combing with the fine split comb, which is then used to accentuate curl around the knots. Continuity of flow must now be secured by the addition of a fine dark line here and there, keeping these short to preserve the bitty character of the wood.

Heart grain may now be suggested by painting in with a pencil overgrainer charged sparingly with turpentine. The latter has an immediate solvent action which opens out the graining colour, forming more lights. These are blended with the hog's-hair softener in a one-way direction across the grain. Then follows a general softening of the remainder, again brushing at right angles to the grain.

Some grainers prefer to omit coarse combing, relying upon the action of a solvent to produce the effect desired. This method should be practised upon graining colours containing varying proportions of oil, turpentine, and pigment, noting the effect obtained during the various stages of setting. Interesting formations are obtained by tools other than brushes. The edge of a feather, a crumpled rag, a coarse piece of sponge, any of these moistened with turpentine and dragged lightly upon the work can be used to advantage in the graining of pollard oak and walnut.

Dapples

These are a prominent feature of quartered oak and must now be wiped out. In this instance the figure is of a much smaller scale and is found in the straighter parts of the grain with the main dapples at right angles to the general direction or flow. Follow this by softening on the lines indicated for quartered oak, and allow to dry.

Overgraining

This is a very necessary detail, for in this wood we find numerous highlights and shades crossing the grain in all directions and, on the whole, vigorously contrasted. A suitable glaze may be prepared from burnt umber and burnt sienna in water, keeping a little of the warmer colour for touching up a few of the knots, and some of the darker colour for painting in a few shades, the work to be finally softened with the 'badger'.

Root of oak

Root of oak is obtained by cutting across the base of the tree where growth is concerned with the development of a fibrous root system rather than with the less compact timber. The grain, as one might expect, is extremely twisted and curled, and, because of its resinous nature, is much darker in tone.

In graining this wood we commence as in pollard oak, by indicating a general sense of direction, again using an old brush, lightly charged with burnt umber. By using the side of the brush in a forceful scraping movement it is not difficult to produce the fibrous appearance of the underlying pattern.

For the next stage we require a sable pencil and a hog's-hair pencil overgrainer or, alternatively, a fan fitch or home-made substitute. With the tool selected we sketch in the broad masses, keeping the brush rather on the dry side. Throughout this operation the lines should be fine and in short lengths, then the knots and parts of the open grain are strengthened, using burnt umber for the whole work. Softening is completed as before, and, when dry, the job is overgrained in water-colour (see plates 4 and 8).

Chapter Eleven

Mahogany; feathered mahogany; rosewood

Mahogany

Ground colour: BS 04.D.44.
Graining colour: Vandyke brown and mahogany lake or burnt sienna.
Tools required: 'Rubbing-in' brushes, mottlers, cutters, badger softener, thin overgrainer and comb, sable pencil and sponges.

This wood is highly valued for its rich, warm colouring, its remarkable smoothness, and the beauty and variety of its figuring. The best examples of Spanish or Cuban mahogany are much finer in colour and pattern than the Honduras variety, which is known as baywood. The latter is much paler in colour, its figure or 'roe' is less pronounced, and its 'mottle' is less interesting.

Both types have their uses in graining; the feather or the mottle of the former is ideal for the treatment of door panels, and the plainer grain as well as the heartwood of the latter variety is suitable for stiles, etc. These four types of grain are easily distinguishable in the two illustrations (see plates 3 and 9).

Honduras mahogany

The grain of this wood is very similar to the smoothly rounded types of American walnut, the main difference being that of scale. Both are lacking in contrast, quiet in effect, finely grained, and displaying the same formation of pores. On the whole, the mahogany exhibits the greater softness and, of course, the larger scale (refer to plates 3 and 9).

Only on rare occasions do we find this type of mahogany finished in its natural colouring; it is usually enriched by the french polishers until some semblance to Spanish mahogany is secured. This is the colour we are expected to match when asked for mahogany, and so long as we can keep within sensible limits and guard against the excessive use of mahogany lake, all will be well.

A truly natural finish requires a ground composed of white and burnt sienna and graining colour of Vandyke brown and burnt sienna, colours which must predominate when matching old mahogany.

Graining colour

The colour for any variety of mahogany can be obtained from the three colours given. Much depends upon the ground colour employed; if sufficiently red, the work may be grained with Vandyke brown alone; if lacking in warmth, the stain can be brightened with mahogany lake. In the present instance we employ only Vandyke brown in the usual water medium.

Graining

This is preceded by the usual surface preparation of fuller's earth; the work is then 'rubbed-in' and, with a 75 mm mottler held as one would employ a one-stroke brush, the preliminary figure is lightly put in. To do this successfully requires colour of the right fluidity, an even pressure upon the mottler, and no variation in the angle at which the mottler is held.

To be more explicit, we commence at the base of the panel, with the edge of the mottler in line with the heart shape; the brush is then dragged upwards for 100 to 125 mm, pulled over to the right, and continued down to the base. The result should be a series of fine lines following a semi-elliptical shape. The whole of the central figure is built up by the simple repetition of this same stroke.

If too much pressure is applied, the left side will be unduly pale and the right side of the figure – which coincides with the downward stroke – will be excessively dark in tone.

Shape is of equal importance; we must preserve some tapering of the heart grain even though it be slight, otherwise a most unnatural and top-heavy pattern will result. The latter effect is frequently the outcome of incorrect brushwork, as, for example, a twisting movement of the brush, causing it to operate compass-fashion when forming the end curves (see fig. 16).

Following the completion of the central shape, the straight side-grain is immediately added, using the mottler at the same angle throughout. When sufficient dexterity is attained, the whole operation of 'rubbing-in' and figuring can be carried out in a few minutes, which leaves ample time for softening.

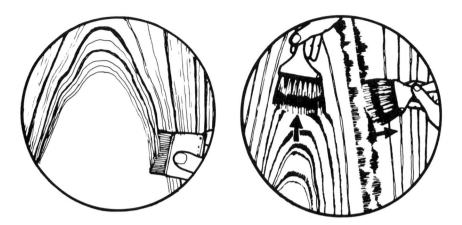

Fig. 16. Use of mottler to produce a tapering elliptical heart grain.

Fig. 17. The various directions of softening mahogany heartgrain.

Softening

This is executed with the badger-hair softener, but in two distinct movements. The heart grain is only slightly softened in one direction – upwards; the side grain is brushed in an up-and-down direction with strokes forming an angle of about 10° with the line.

The slanting strokes are applied from each side in turn so that the lines are actually softened by strokes conforming to the direction of the strokes in the capital X (see fig. 17).

When the first graining is dry, we take a sable pencil and with colour a shade darker, we strengthen the end grain here and there, keeping the lines widely spaced as in American walnut; the work is softened in an upward direction.

Overgraining is carried out in an oil medium, using a thin wash of Vandyke brown slightly tinted with mahogany lake. The mottle is applied to the side-grain and the whole is lightly stippled with the badger softener. This gives a good suggestion of the fine pores.

Alternative methods

(1) 'Rub-in' and stipple in water stain; grain in an oil medium; overgrain and mottle in water.
(2) 'Rub-in', mottle and stipple in oil; glaze and grain in watercolour.

Feathered mahogany

The term aptly describes the intricate figuring of this highly decorative wood which owes its pattern to the fact that it is cut from the topmost part of the tree trunk through the actual base of a main branch. The feather is also to be found in other trees, some being identical in figure to genuine mahogany (see fig. 18).

For the graining of this wood we use Vandyke brown with a touch of mahogany lake; some grainers prefer mahogany lake and black, others employ lake and burnt sienna, but nearly all are unanimous in the choice of a water medium. The chief tools required for the first stage of the work are: a small sponge, badger softener, a 40 mm camel hair cutter or a mottler of similar size.

Graining

'Rub in' with watercolour of medium tone; then, with a sash tool loaded with dark colour, paint in the broad central shape, ascending with numerous outward curves somewhat like the rising jet of a fountain. The whole work must be carried out as wet as possible, otherwise the colour may dry before completion; furthermore, the silky-soft effect can only be obtained by the manipulation of wet, not semi-dry, colours (see fig. 19).

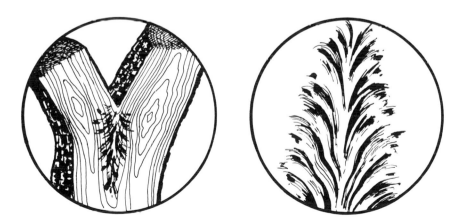

Fig. 18. View through the base of a main branch showing the source of feathered mahogany.

Fig. 19. The form of feathered mahogany.

Immediately following the application of this groundwork we take a small, damp sponge and wipe out the lights which form the main pattern. These shapes might be described as similar to water falling in a downward curve from numerous ascending jets, the whole forming an arched cone. The sponge must be rinsed very frequently to maintain cleanliness of colour, and the wiping-out movement should always follow a downward direction, working on both sides alternately (refer to plate 9).

The figuring on either side should be very close, in fact, the lights may slightly overlap to prevent excessive width at the base; then, as we near the edge of the panel, the formation can be more widely spaced and the lines more broken in character.

Mottling

This is carried out upon the wet colour; this will be fine, close, and most pronounced in the straight grain and upper parts of the feather. In the latter case the mottle shows a pronounced curve as it sweeps gracefully outwards from the centre. The whole is now softened by light strokes in all directions, and then the cutter is employed to take out the sharp highlights which intermingle with the curved mottle: these are softened vertically and the work is allowed to dry.

We must now use a thin overgrainer, lightly charged with a very dilute colour and split by combing; with this tool we overgrain the whole of the work, always following the direction set by the underlying shapes, finally softening upwards and outwards with the 'badger'.

Overgraining

This requires but a thin glaze of oil stain enriched by a touch of mahogany lake.

Following its application, the highlights and parts of the mottling may be wiped out; the whole is then lightly stippled. When the suggestion of pores is lacking in strength, we may await the setting of the glaze and then apply a little darker colour with the tip of the badger softener.

Alternative method

This is particularly useful during periods of hot weather and (which is likely to be of value to the beginner) may be commenced by glazing and stippling in watercolour, graining in an oil medium, and finished by glazing in water.

During the graining process the sponge is replaced by a double thickness of rag held over the edge of the thumb, mottle is formed by the stiff hog's-hair mottler, and the fine high-lights are obtained with a veining horn. A flat fitch is also useful for strengthening parts of the grain.

Overgraining is not attempted until the stain is well set. Afterwards it may be safely carried out in thin oil colour, softening as before.

Glazing in this case will necessitate a water medium.

Rosewood

Ground colour: BS 06.E.56.
Graining colour: Vandyke brown, mahogany lake, and black.
Tools required: 'Rubbing-in' brushes, sponge, flat fitch, sable pencil, thin overgrainer, and badger softener.

It is not easy to give a brief but adequate description of all the characteristics of this wood. In colour it exhibits the deeper tones of mahogany, interspersed with numerous dark lines and broad streaky patches of varying tones.

The grain shows a definite sense of direction and yet is extremely complicated by the inconsequential arrangement of adjoining bands of figuring. Heart grain dominates the whole; this varies from coarse to fine, sometimes running in an upward direction, at other times reversed. A dark sketchy line frequently divides the varying formations, and the general impression is one of softness and richness (see plates 4 and 8).

Graining

May be carried out in water or oil colour, the order of working being the same in each case. For our present purpose we shall describe a water-colour process which appears to be more widely employed.

The surface is 'rubbed-in' with a medium tone of Vandyke brown and followed immediately by the formation – by scraping with the side of the brush – of several lighter strips of colour. With the fitch and a darker tone of Vandyke, strengthen and sharpen the edges of each dark strip, then, using the brush as a drag, indicate the general direction of heart and straight grain. Blend with the badger softener and allow to dry.

Stage two is concerned with additions to, as well as the strengthening of, the preliminary grain. For this we employ a mixture of Vandyke and black, or mahogany lake and black, again in a water medium. Some of the work may be fine in detail, but an effect of broadness is desirable when putting in the heart grain; soften in the direction of the grain.

Overgraining

The use of oil colour enables the work to be enriched in colour and strengthened where necessary. The glaze colour is made up from mahogany lake and black, with a deeper tone for accentuating parts of the figuring. The glaze is applied, lightly flogged with the 'rubbing-in' brush, fine grain is put in with the thin overgrainer, and strengthened with the sable pencil, the whole being completed by slight softening and flogging with the 'rubbing-in' brush.

Chapter Twelve

Maple, satinwood, and pine

Maple

Ground colour: BS 08.C.31.
Graining colour: Raw umber and raw sienna.
Tools required: 'Rubbing-in' brush, mottler, cutter, badger softener, bird's eye dotter, and crayon.

There are several types of this beautiful silvery wood, all of which are remarkable for their silky mottle and general cleanliness of effect. Examples vary in colour from the usual 'cream' tints of new wood to the stronger yellowish tones of the old timber; other varieties – equally important to the grainer – are almost silvery grey (see plates 4 and 6).

The pattern, too, will differ according to the species; also the manner in which the tree is cut. Generally, the outer part gives the 'bird's-eye' maple, while the heartwood produces the plainer 'curly' maple'.

Bird's-eye maple

It is usual and more expeditious to represent this wood entirely in a water medium. The ground is prepared by degreasing, after which the work is 'rubbed-in' with water very slightly tinted with raw umber and a touch of raw sienna. This colour can hardly be described as a stain; it is, in effect, just tinted water. The point is emphasised because of its importance in the securing of clean results.

Figuring

This must be worked in at high speed upon a very wet coating; it is also helpful to employ a large mottler, as this fills the space in a very short time.

The immediate object is to cover the whole panel with close mottling, with lines intersecting at all angles, but not far removed from the horizontal. Soften lightly by 'badgering' in a horizontal direction.

Many devices have been employed for putting in the bird's-eyes. Some, such as the fingertips, rough cork, or a small sponge, provide an extremely rapid means of introducing dots, but are most unlike the natural 'eyes'. The latter are defined by a dark outer edge and a pale centre, rather like a horseshoe, a shape most easily reproduced by any simple printing device.

Apart from the 'maple-eye dotter' previously mentioned, there are many improvised tools which are quite crude, but efficient. These include circlets of rubber, chamois leather, thin felt or other stout material, small pieces of potato, and others too numerous to mention. The shape desired is an incomplete oval, and so long as this is obtained, we can continue on the right lines.

The spacing or grouping of the 'eyes' is subject to few rules. These are often closely arranged and at other times widely separated. They do, however, occur along the points of junction between lights and darks, and are placed with the break uppermost to suggest highlights and shadow.

These are all put in with a tint of burnt sienna and lightly softened. Then, whilst the mottling is still wet, we take a small cutter and form numerous highlights extending sideways from the various eyes. Soften these in a horizontal direction.

Overgraining

The best tool for this work is undoubtedly a crayon-pencil, approximately burnt sienna in colour. With this we sketch in the centre grain which takes a formation similar to curly pitch pine, but small in scale.

Throughout this process the lines must never run through any 'eyes'; they may touch, or curl around the 'eyes' just as they curl when crossing the mottled parts, and, as in pitch pine, the shapes set by the heart grain must be followed by the side-grain until, as a result of interference by other angles of the mottling or by 'bird's-eyes', the shape must be changed. The work is ready for varnishing as soon as overgraining is completed (see fig. 20).

Oil-colour method

This medium is extremely useful when graining bird's-eye maple in the

Plate 9

Mahogany (heartwood)

Feathered mahogany

Plate 10

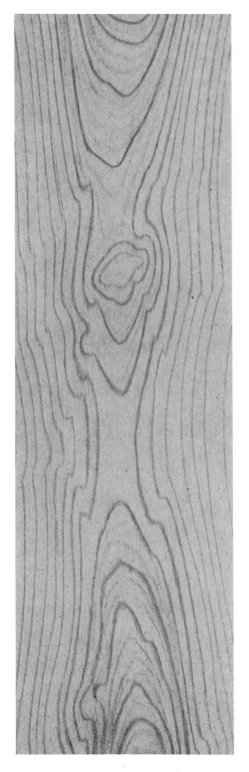 

American walnut   Burr walnut

Plate 11

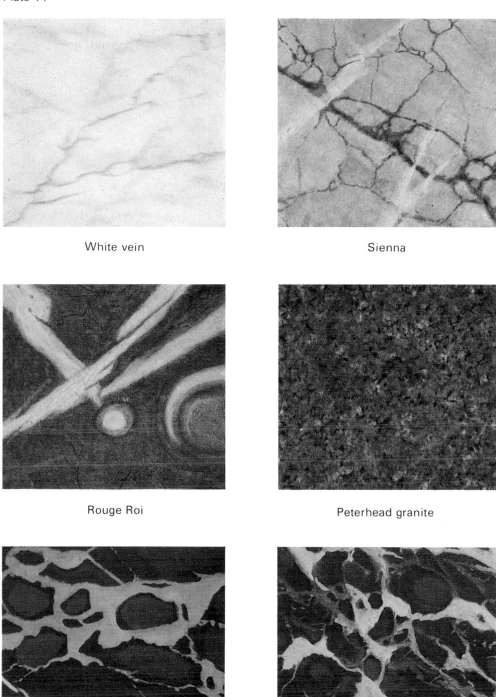

White vein

Sienna

Rouge Roi

Peterhead granite

Black and gold

St. Anne's

Plate 12

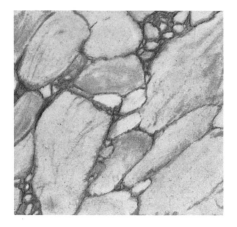

Breche violette

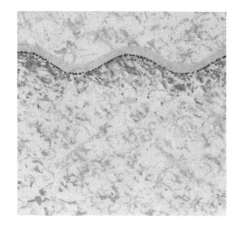

Sponge stipple (distemper)

Cipollino

Sponge stipple (oil colour)

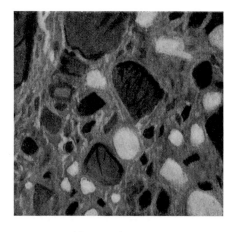

Vert antique

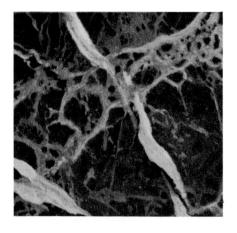

Vert de mer

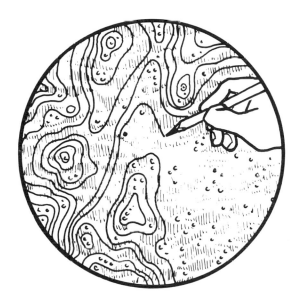

Fig. 20. Use of crayon to produce the grain of bird's-eye maple.

heat of summer. There is little change in the method; an oil medium can, in fact, be of great assistance, because it gives the grainer-student more time for each stage of the work. The mottling can be put in with the 'rubbing-in' brush and supplemented by finer work with an ordinary mottler, but considerable pressure will be required as the colour begins to set.

In all forms of curly mottling it is usual to press the fingers well into the bristles of the mottler, near the tips. Not only will this produce the required zig-zag shape, but the brush will be stiffened, and thus lift out the colour more efficiently.

When the mottling is partly set we may add the bird's-eyes as before, wipe out the highlights, and put in the overgrain with either crayon or oil stain and a fine brush. Finally, soften lightly with a hog's-hair softener.

Curly maple

This is ideal for the treatment of stiles, mouldings, etc., its comparative simplicity providing suitable contrast to the more ornate 'bird's-eye' panels. We shall describe the graining of this wood in water, but it can, of course, be grained equally well in oil.

The panel is 'rubbed-in' and mottled to give a less compact and more horizontal series of lines, but keeping the centre much plainer. Soften with the badger softener, first in a horizontal direction and then with a few vertical strokes.

Overgraining

Put in with a crayon as before, the heart shape being larger in scale and similar in form to a simplified type of Hungarian ash.

Satinwood

Ground colour: BS 10.C.31.
Graining colour: Raw and burnt sienna, and Vandyke brown.
Tools required: As for mahogany.

This wood is well named, its lustrous satin-like sheen distinguishing it from all others and ensuring its recognition as one of the most decorative of the light-coloured woods.

Its grain might well – apart from colour – pass for mahogany, which it closely resembles. Not only do we find the plain heartwood, but also the mottled and feathered varieties, all of which are grained in exactly the same manner as mahogany (refer to plates 5 and 7).

The mottled variety is universally employed in the making of furniture, and it is this type, together with heart grain, which will give an adequate grounding in the use of the colours employed.

The stain is prepared from raw sienna with a slight touch of Vandyke brown, well mixed in a water medium. After 'rubbing-in', we employ a small sponge to take out a series of short lines in the parts to be mottled, but leaving a space quite plain for the heartwood. A small cutter is now used to mottle upon and extend the light patches in a horizontal direction so as to overlap upon the darker parts. Put in a few widely spaced lights to form a background for the heart, and then soften horizontally with the badger softener.

The heart shape and the overgraining is painted in with raw sienna tinted with burnt sienna to form a slight contrast with the underwork. Soften as in mahogany.

Pine

Ground colour: BS 08.C.31.
Graining colour: Raw umber, raw and burnt sienna, black.
Tools required: As for pitch pine.

Although this timber is comparatively cheap, it can, in the right surroundings, appear fitting and attractive. It displays a certain cleanliness of colour and subdued grain, qualities which are likely to ensure its wider use by grainers.

Graining colour

Prepared mainly from raw umber with a touch of raw sienna, but this can be varied by adding a little black for those parts streaked with cooler tones. Only the slightest of colouring is necessary, and the medium may be water or oil, as desired.

The work may be 'rubbed-in' with watercolour and then figured in either water or oil colour, or the whole job may be executed in oil colour. The main difference, as far as the figuring is concerned, being that watercolour gives a sharper and more elongated series of end shapes, while oil colour is finer and more controllable (see plates 4 and 6).

For the figure we require a little of the 'rubbing-in' colour tinted with burnt sienna. If watercolour is employed, it must be stirred frequently to maintain an even colour during the painting of the heart grain. The brush should not be heavily charged with stain or the effect will be harsh. It will, however, be advantageous to apply a little extra colour at the curved ends.

Softening is performed as the work proceeds; the 'badger' being used in a one-way direction towards the curved ends. This draws up the extra colour on the curved ends producing a hard edge on the top and a softened effect beneath (see fig. 21). Finally, the plain side-grain is added with the split, thin overgrainer. Knots are more realistic than decorative and might well be left out of the work, but one here and there could be painted in with a dilute wash of burnt sienna.

Overgraining is unnecessary.

Fig. 21. Softening through the centre of pine heartwood.

Chapter Thirteen

Walnut

Ground colour: BS 06.C.37.
Graining colour: Vandyke brown and black.
Tools required: 'Rubbing-in' brush, sponge, fitch, sable pencil, thin
  overgrainer, mottler, badger softener, and a few large feathers.

In this chapter we deal with varieties other than American walnut, varieties which, by their contrast of tone and pattern, are full of interest and decorative possibilities. Italian walnut is regarded as superior to others, but interesting figure is not peculiar to any one species; it is a product governed by local conditions of growth, by the manner of cutting, and in some cases by the development of burr or abnormal growth on the side of the tree.

There is a quiet richness about walnut which is gained by depth rather than by brightness of colour. One sees below the surface, as it were, and in most varieties we can trace a soft mottle upon which is a delicate under-grain, then a darker and more vigorous pattern, apparently unrelated to the first, but which lies below the glazed surface and its numerous fine pores.

In the following pages we shall explain the methods employed in the representation of three distinct types of walnut. The methods are arbitrary and in consequence may be interchanged as desired.

English and French walnut

We purposely omit the reproduction of burr in this section of the work, but aim at the production of a type which is plainer in character and which provides the necessary experience in the peculiar technique involved in the more elaborate examples (see plates 7 and 13).

The surface should be prepared and 'rubbed-in' with a thin glaze of

Vandyke brown in water medium. Put in plenty of fine mottle at right angles to the intended direction of the finished grain, and finish with the badger until the mottle is just discernible. It will, of course, be appreciated that additional warmth may be secured by the use of burnt sienna, and that the addition of black will give a cool tone.

In the second stage of the work we put in a very delicate under-grain, using the glaze colour and a thin overgrainer, lightly charged, and split by combing. The badger softener is in constant use, softening across the grain in one direction, but taking care not to smudge or brush the lines one into another.

Considerable practice will be necessary before this technique is properly mastered: it needs the right amount of colour and a half-brushing, half-stippling movement, with the tips of the softener pushing, rather than dragging, the colour forwards. 'Strap down' before continuing with the next process.

Surface grain is now painted in and softened. The colour may be much deeper in tone, and for those parts requiring strong contrast may contain a little black. Most of the grain is applied with the overgrainer as before, then, when the whole area is covered, strengthen parts of the grain with dark colour applied with the sable pencil.

At some time there may be 'cissing', i.e. difficulty in getting a smooth continuity of line. When this occurs, it is necessary to breathe upon the grained work, but if this is of no avail, dab the whole surface with a damp chamois leather.

Overgraining constitutes the last stage of the work. This is done in oil and requires only a thin glaze, which may be stippled or flogged to give a suggestion of pores.

Italian walnut

This variety is characterised by its curly formation, close and well-defined mottle, and its broad bands of contrasting tone. Most of the grain, including the knots, is displayed in the darker areas, the lighter parts being comparatively plain (see plates 7 and 13).

The work is usually spread over two stages, the first being to 'rub-in' with Vandyke brown and a little black in water medium, then, with a sponge, take out the light strips. The wet colour is now mottled with a 100 mm brush, keeping the angle fairly constant in some parts, and more varied in positions to be occupied by rings and knots. The work is immediately softened.

When dry we use an overgrainer and sable pencil to add further detail, keeping pale tones for the lighter parts and strong tones for other sections of the work. Soften with the badger, and, when dry, glaze and flog in an oil medium.

Burr walnut

To grain this wood as it is some times seen in furniture, i.e. all burr, is too lengthy a process; nevertheless, the grainer is often called upon to reproduce a more simplified version upon door panels and other important centres of interest.

It is usual in such cases to combine the simple type of walnut with clusters of burr, the former curling around and linking up the groups of knots in a natural manner. The contrast provides additional interest, gives a sense of direction, and is altogether preferable to either when used alone (see plates 7 and 10).

In the graining of this example we shall employ both oil and water colour, with the simplest kit of tools.

The graining colour is Vandyke brown in a water medium; this should conform to a medium tone of the finished work. The surface is 'rubbed-in', and with the edge of a feather we sketch the plain parts of the grain, adding several loops and sharp angles as we work from the upper to the lower part of the panel.

The feather grips the wet surface and requires firm pressure to drag it through the graining colour, but the result is quite unlike that obtained by any other tool. The veining produced in this manner should be extended over the greater part of the area, leaving empty spaces in the parts to be filled by burr. The veins are now softened in one direction, always using the badger at right angles to the grain, and taking care not to smudge the lines (see fig. 22).

The accidental effects gained in this manner are sufficient to ensure variety of form but not of colour. The latter quality may be obtained by the partial removal of graining colour by means of a mottler or sponge as the feather graining proceeds. Each light strip is figured in a rather plain manner, using a small feather for such parts.

At this stage, we put in a background for the burr, using a sponge moistened with the stain and applied by stippling. The flat fitch is then used with a twisting motion to open out several larger patches, after which the work is lightly softened in all directions.

We now prepare a little dark colour by mixing black and Vandyke brown, then, with the pencil, proceed with the strengthening of several of the darker veins, softening as before. The fine outlines of the burr may now be pencilled in with varying tones of colour. We are not without guidance in the placing of these circular shapes, for the shadowy background should be full of suggestions which await further development. The softening of these must not be overdone; a combined stippling and softening movement is quite sufficient to remove harshness without spreading the colour unduly.

The work is glazed in oil colour, which permits minor additions to the figuring as well as the introduction of mottling; this is followed by softening and completed by stippling to produce pore-marks.

Fig. 22. The use of a large feather to apply colour. This technique is applicable to both graining and marbling.

Chapter Fourteen

Quartering and inlay effects

The treatment of large surfaces is often something of a problem, and it is with a view to finding a solution, or at least to suggest simple methods by which such areas can be broken up, that this chapter is directed. The decorative value of quartering requires no comment here, except to point out that it is well within the capacity of any grainer to reproduce the effect in paint, and to do it economically.

Opportunities for this type of work or for the representation of inlaid woods, or a combination of both, are more plentiful today than ever before. We have flush doors, signboards, the panelling in shop windows and other surfaces, which provide scope for more decorative treatment.

Quartering

This natural outcome of veneering is always in demand. In the case of figured woods we usually find that two or four slices have been cut, one after the other, from the same piece of timber, in order to secure examples of almost identical pattern for use side by side, or quarterly. Considerable ingenuity is shown in the arrangement and fixing of the pieces to obtain suitable contrast of colour and pattern.

A good eye for pattern is equally important when suggesting such arrangements in paint, and it is always advisable to prepare a rough sketch, and thus obtain a preview of the design before commencing upon an actual job. The preliminary pattern may be sketched upon a painted surface with a graphite pencil, or a pounce can be made of the leading lines of one quarter, and transferred by dusting with dry colour of suitable tone.

The latter method is accurate and reasonably quick – especially for quartering. By centring a door horizontally and vertically and ruling in the pencil lines, the initial shapes are obtained. The first quarter is pounced and the pattern turned over, sideways, for the next, then turned downwards for the third, and again sideways to complete the job.

If the dry colour is tied up in muslin and the holes in the pounce are fairly large, it will require no more than a few seconds to pounce each quarter. The dotted lines can then be sketched in with lead pencil.

This type of work may be completed by working along the following lines:

(1) Rule the centre line, then apply a thin glaze of watercolour, and mottle and soften.
(2) With a thin overgrainer and the same colour, put in the preliminary veining, softening as the work proceeds. It is, of course, necessary to grain the left half first, and then match this when figuring the right half.
(3) Strengthen the grain with darker colour (Vandyke and black), using a fitch and sable pencil.
(4) Fix with a coating of thin oil glaze, touch up the burr, and stipple the whole.

This, when varnished, would complete the ordinary type of job, but to carry the work a stage further and include a band of Macassar ebony, it is wise to employ a gloss varnish. This, when hard, is rubbed down with fine wet or dry paper until the surface is smooth and matt.

For the inlay it is necessary to mark out the border in lead pencil and then grain in Vandyke brown and a touch of black, again in a water medium. The graining must be extended beyond the pencil lines, but the latter should be faintly discernible through the lighter parts of the grain.

When dry, the whole of the border is fixed by a coat of varnish and turpentine. This operation calls for great care, for any part of the watercolour becomes fixed when touched with varnish. Good brushwork is particularly essential at the corners and outer edges of the border.

Surplus colour can be sponged off when the varnish is dry, and the whole surface completed by varnishing.

To quarter a door in straight-grained mahogany or other wood requiring a water medium is a comparatively simple operation. The top left and bottom right-hand sections are grained and fixed with varnish, then, on the following day, surplus colour is sponged off and the two remaining sections are grained and fixed. The work is cleaned up and finished on the third day.

It is usual, when quartering, to show some slight variation in tone between adjoining sections of the work. In natural wood this effect is due to reflected light, and at times is very pronounced.

These tone variations are helpful in another direction; they assist in the production of sharp definition between the quarters. It is, however, of much greater importance to be accurate when 'cutting-in' with the fixing varnish. The brush must be sparingly charged in order to prevent varnish from spreading across the pencil lines. It is neatness in this part of the process which makes or mars the whole effect.

It should scarcely be necessary to add that the angle of slope which is set in the first quarter must be strictly adhered to in the remaining sections.

Quartering is sometimes used in conjunction with a central elliptical panel, in which the feather or heart grain is shown. In such cases it is better to grain and fix this centre panel before adding the quartering. The preliminary setting out of the ellipse should be done by means of a trammel, not by the gardener's method.

Quartering in oil colour

Also a two-day process, the quarters being marked out as before, and the sections grained in the same order. For this work it is better to employ several 25 mm strips of masking tape to protect adjoining quarters and preserve a straight edge whilst graining. These can be removed immediately after graining. On the second day, the tape is again employed to protect the grained parts until the two remaining quarters are figured.

Inlay effects

The suggestion of inlaid work is mainly a watercolour process, this being the thinner of the two graining media. Oil colour has its uses, but these are limited to the production of dark graining upon a lighter undergrain, such work being executed with the aid of masking tape or other protective covering.

To display several woods upon one surface it is usual to grain the whole area to represent the lightest of the woods, this work to be fixed by the usual coat of varnish and turpentine.

This is followed by graining the whole area in the next deeper tone. The desired shape being traced upon its dry surface will enable the part to be fixed. The dark-toned wood is grained on the third day, its pattern is traced and carefully fixed as before, and on the fourth day surplus colour is sponged off, and the whole surface varnished.

The formation of delicate pattern in pale wood upon a dark ground can be achieved by another method.

The whole surface is grained in the paler tone and fixed by varnishing; then the pattern to be finished in this wood is traced upon the dry surface and painted in with Brunswick black.

When dry (in about one hour) the work is grained in watercolour to represent the darker wood; this, when dry, is soaked with turpentine and gently brushed with a camel-hair mop. The black is easily removed without causing injury to the superimposed graining, and the work may then be varnished. By this method there is no apparent difference in thickness between the two varieties of graining.

Chapter Fifteen

Marbling

Marble, as distinct from granite, is a general term applied to any form of compact stone which is capable of giving a polished surface. The true marbles are rock limestone, crystallised by pressure and intense heat, and coloured by the percolation of mineral or other foreign substances.

A thousand years is a mere nothing in the formation of marble, and, in consequence, we find fossilised fragments of animal and vegetable matter interposed within the substance of many species. Fissures, due to movement of the earth's crust, have allowed the infiltration of coloured deposits which, under pressure, have solidified.

Although accidentalism is responsible for much of the colouring and formation of marble, and, unlike timber, it is not related by a common order of growth, each variety is clearly distinguishable by some characteristic imparted as a result of the circumstances governing its origin.

Laminated marble is easily recognisable as one formed in the first place by layer upon layer of coloured silt, e.g., 'Cipollino'.

Brecciated marbles are a conglomerate mass of small angular fragments cemented together by lime or other mineral substance. These are the result of violent fragmentation of the original bed, the various pieces being scattered and afterwards re-set, e.g., 'vert antique'.

Crinoidal marbles are those composed almost entirely of shell fragments, which give them a speckled appearance, e.g., 'Roman Stone'.

Stalagmitic beds, caused by deposits of carbonate of lime, petrified in the course of centuries, produce 'Roman Travertine' and 'Onyx'.

Serpentinuous may describe the undulating form of veining, but in marble it indicates the mineral serpentine which forms the basis of a number of green marbles, e.g., 'Tinos', and the red variety of 'Levanto Rosso'.

Variegated marbles are those with textured or clouded effects broken by sinuous veining, e.g., 'Sienna'.

A study of the character of the various formations is quite as important as that of colour, one being complementary to the other. The marbler should miss no opportunity of examining as many as possible of the larger slabs, for it is here that characteristics are more strikingly apparent. At the same time, numerous sketches and colour notes can be made for future use.

Surface preparation

The majority of marbles may be represented upon a broken white ground of egg-shell gloss finish. The surface, for obvious reasons, should be exceptionally well prepared so as to present a high degree of smoothness, representative of the polished surface of marble.

Tools and materials

No elaborate kit of tools is required for marbling, although it will be necessary to obtain a good 'badger' and also a hog's-hair softener of about 100 mm in width. These are the only expensive items; for the remainder, it is quite sufficient to collect the following: 'rubbing-in' brushes, fitches, sable pencils, rags, goose-wing feathers, natural sponges, palette, palette knife, and crayons.

Pigments

Should be as fine and pure as possible, and ground in oil. It frequently happens that opaque pigments have to be employed in the representation of certain marbles, and in such cases it is advisable to dilute the colour until the consistency is a little more than a thin wash. As a general rule pigments are selected for their natural translucency, a quality which is even more important in marbling than in graining. The transparency, depth, and perfect flatness of marble can best be suggested by glaze colours which, if properly manipulated, will also simplify the production of the characteristic stony appearance.

Medium

For general use this is a mixture of two parts turpentine and one part refined linseed oil, with the requisite amount of liquid oil driers. This medium is also known as 'gilp' and may be employed for the preparatory 'oiling-in', or moistening of the ground, and also for the universal thinning of pigments. There will be special cases where the addition of a pale oil varnish, or the use of varnish and turpentine, is preferable.

Watercolour may also be exploited in marbling – as in graining – to expedite work and produce effects not easily obtainable in oil. It is often employed in the preliminary stages, and the work completed in an oil medium, with perhaps a final glaze in watercolour. This combination of oil and water processes offers a wide field for experimental work of undoubted value.

Crayons

These may be used in conjunction with oil colour and are particularly useful when putting in the fine sketchy veins of sienna etc.

Grounds and colours employed

Type of marble	Ground colour	Pigments used
White vein	White	White, black, ultramarine blue, yellow ochre
Sienna	White	Raw and burnt sienna, middle chrome, zinc white, ultramarine blue, Indian red
Breche violette	White	Indian red, crimson lake, ultramarine blue, ochre, white, black
Cipollino	White	Prussian blue, raw sienna, mid Brunswick green, black, white
Granite	White	Venetian red, burnt ochre, burnt umber, black, white, ultramarine blue
Rouge Roi	White	Indian and Venetian reds, burnt sienna, ochre, ultramarine blue, black, white
Black and gold	Black	Raw and burnt sienna, middle chrome, zinc white
St. Anne's	Black	Zinc white
Vert de mer	Black	Prussian blue, raw sienna, Brunswick green (middle tone), white, black
Vert antique	Black	Middle Brunswick green, black, white

It should be noted that the grounds and colours suggested above are those used by the writer in the execution of samples illustrated in the following chapters. Some are universally employed, others are often varied according to the individual taste or method adopted by the marbler. Such freedom of expression is more desirable than hard-and-fast rules, as the beginner will appreciate when working out the various exercises. The experience gained from the representation of this selection should enable the student to reproduce other varieties of marble without difficulty.

Chapter Sixteen

White-vein marble, sienna, and breche violette

White vein

White marble is found in all parts of the world and in many varieties which range between pure white or statuary, without spots or veins, to the clouded white and grey, intersected by veins of darker grey. It is the latter species which is employed in the decoration of buildings and which therefore provides copy for our immediate purpose (see plates 11 and 17).

The first stage in the working of this marble is that of clouding or scumbling the ground. For this we require thin white undercoat as a 'rubbing-in' colour, and small quantities of light and medium-toned grey. The latter may be mixed upon a palette board using white, black, and a touch of ultramarine blue. The light grey should be neutralised by the addition of a slight trace of ochre.

Proceed with the 'rubbing-in', using a large brush, then apply irregular patches of light grey, using a large fitch in a fairly dry condition. With a little of the dark grey strengthen one or two of the tones already applied and blend with a hog's-hair softener, using both horizontal and vertical strokes until a faint cloudy effect is obtained.

Veining is introduced before the groundwork becomes set: the finer veins are put in with a translucent wash of light grey and lightly softened, then the darker lines are added by means of a fine sable pencil lightly charged with dark grey diluted with oil medium.

These are the principal veins which must establish a definite sense of direction. The formation is based upon sketchy but fairly straight lines, branching and intermittent in places, but never developing into pronounced curves. Softening is carried out in one direction only and following the general run of the veining. By this means we secure the effect of one soft and one harder edge as displayed in the natural marble.

The final stage, that of glazing, must not be attempted until the underwork is hard and dry. This process consists of lightly glazing the whole surface with a thin wash of white, then, with a fitch and white of denser quality, we add the numerous spots and patches which appear to lie upon the marble surface between the veins. The work is completed by softening in all directions, and, when dry, may be varnished with a pale varnish.

Some marblers prefer to use a glaze of white gloss paint and turpentine to replace the glazing and varnishing. In such cases the white patches are introduced either by painting upon the partly set underwork or by wiping out. The gloss paint gives a whiter finish and maintains its purity of colour for many years.

Sienna

This well-known type of variegated yellow marble is named after the Italian province in which it is quarried. Throughout the ages it has been highly esteemed for its richness, warmth, variety of colouring, and harmonious relationship with other marbles. It is not uncommon to find tints of pale cream, rosy red, grey, and deep yellow, broken by veins of black, purple, reddish-grey, and brown, all softly blended, and forming a pleasant harmony of sequence with the predominant yellow ground.

Some examples are pale in colour and only slightly broken with soft grey veins interspersed with white veins of a transparent spar-like appearance. Others display masses of small stone-like shapes, linked by fine, sketchy veins. Bold veining is also found interspersed among large and small slabs alike, which fact may be used to advantage in giving the work a centre of interest and a sense of direction (refer to plates 11 and 14).

Beginners invariably experience some difficulty when reproducing the veining of this marble, and it requires constant care and deliberate effort to avoid the mechanical repetition of stones or slabs of one particular shape or size.

Freedom and variety are best secured by the use of short broken lines of varying colour and thickness, the thin lines being put in with the tip of the writing pencil and the soft broad parts being produced by a dragging and rolling stroke with the side of the brush (see fig. 23).

Notice the disposition and grouping of large and small areas, the latter forming compact masses but never becoming chain-like in appearance. Avoid the formation of true squares and circles; there are many rounded shapes, but these are more cloud-like than circular. Finally, be on the look-out for accidental effects or shapes which are

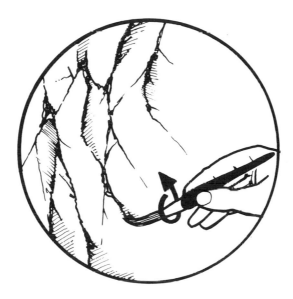

Fig. 23. Rolling and dragging a sable writer to produce irregular veining.

always present and which suggest natural divisions awaiting development.

Method

Charge the palette with the colours arranged in their correct tone order i.e., ranging from white at one end to blue at the other. These should be mixed with a little of the oil medium – if too stiff – and two dippers, one containing the medium, the other turpentine, may be fixed to one edge of the board.

Before the application of colour it will be necessary to 'oil-in' the ground. This consists of moistening the surface with some of the medium, applied with a clean rag, with the object of facilitating the blending of superimposed colour washes. Three or four 25 mm brushes or fitches will be sufficient for the preliminary clouding of the ground, but all tools and equipment must be quite clean and free from dust.

Dip each brush into the medium prior to picking up the various pigments from the palette; this should ensure that each colour is diluted sufficiently to give a translucent tint or scumble, which may be varied in tone according to the amount applied and the extent of its spreading.

Patches of raw sienna will be universally spread over the surface, with touches of chrome, burnt sienna, and grey-violet distributed throughout. Deep tones give an impression of weight, and should therefore dominate in the lower parts of a panel, and, for the same reason, pale tones may well occupy the upper parts. A mixture of Indian red and ultramarine blue gives a suitable grey-violet, but owing to the possibility of this combining with yellow and forming a green, it is only applied upon the paler areas.

After the preliminary scumbling or clouding, the whole of the white ground will be more or less obscured by irregular washes of colour, which must be lightly blended with a soft hog's-hair softener and then broken up by stippling with a coarse sponge or crumpled rag, or paper, moistened in turpentine. Within a few minutes the solvent action of the turpentine will cause the scumbling to open out: this may be checked by dabbing with dry cloth, crumpled as before. The work is then softened in all directions.

At this stage the ground may be faintly scumbled in all directions, with thin washes of white, pale grey, and yellowish tints. These are applied by means of a feather, the ruffled edge producing lines varying from extreme fineness to coarser and more intermittent shapes. Some may be run out with turpentine only, after which the whole is again softened. This may seem to be a rather complicated process, yet the whole panel can be oiled, scumbled, textured, and be ready for veining within five minutes.

The large veins are put in first, with a mixture of Indian red and burnt sienna, using varying proportions and sometimes adding a little blue. It is a good plan to charge one side of the pencil with brown and the other with blue-violet, to produce colour changes as the brush tip is dragged and rolled across the work. These lines should be sketchy and irregular and should follow the edges rather than cross over the scumbled patches of colour. Soften off before continuing with the addition of subsidiary veins.

Secondary veins are finer and paler in colour. For these we use varied tones of reddish- to yellowish-browns prepared from raw and burnt sienna. The areas to be enclosed by these lines are often suggested by the underlying pattern; it is merely a question of controlling and utilising the accidentals which already exist, and at the same time avoiding the production of acute angular shapes.

The finest and darkest lines are now added to sharpen the edges or centres of the principal veins. The point of a No. 1 pencil will do the work satisfactorily if kept rather sparingly charged with blue-violet. Very little softening is required; in fact, stippling is often effective as a means of toning undue harshness of secondary veining.

It is now advisable to introduce some variety of tone by wiping out and thereby lightening several of the small stones and parts of the larger ones. Clean rag, moistened with turpentine, offers the best and most easily controlled tool for the job. The first stage is then completed by the addition of a few white veins, put in with a feather charged with white undercoat, thinned almost to the point of transparency.

When dry, the work is glazed with transparent washes of the colours previously used. During this process there is an opportunity of strengthening the colours and tone contrasts and thereby increasing definition between areas flanking the principal veins, and also – and to a lesser extent – between the adjoining stone-like shapes. In the latter case it is only necessary to touch up some of the outer edges along the veins.

Should the work appear too colourful after the first stage, the whole should be glazed with a wash of raw sienna. This would at least ensure unity between the various parts and colours.

Breche violette

The quarries which produced this most decorative marble are situated in the Carrara Mountains in Italy, and although these have long been exhausted, there remain in many of our public buildings examples which can hardly fail to compel the attention of future generations of marblers.

This is one of the many types of breche marble and is particularly attractive on account of its warm, clean colouring and interesting pattern. The main colour is creamy white, clouded with soft tones of blue-grey, and interspersed with veins of almost all the spectrum colours. The principal veining is mainly of blue-violet, which constantly changes in tone, and which encloses irregular and angular shapes of white, variegated, and shaded patches.

The method of working is very similar to that employed for sienna. The ground is 'oiled-in' and variegated by scumbling with thin washes of greyish-pink; made from white, crimson lake, and a touch of ochre, with pale blue-grey, made up from white, black, and ultramarine blue; and with ochre alone. The colours may be applied by means of large feathers or by pieces of crumpled rag, the various tints being unevenly distributed over the whole surface.

Tints of the deeper tones of red, purple, and violet-greys are now prepared from Indian red, ultramarine blue, lake, and black. These are considerably diluted with the medium and scumbled over those parts which are to be more strongly veined. The whole is softened prior to veining.

The principal veins are painted in with varying tones and hues of grey-violet, prepared from black, ultramarine, and Indian red. It is a good plan to use a one-stroke sable writer for this purpose, charging each edge with a slightly different tone. By sketching with one end of the brush, a thin line is obtained. Then the whole of the width is gradually brought into use to form a long triangular shape, after which the other edge is made to continue with a fine line until another change of thickness is desired. This continual rocking movement of the brush produces the variety of shapes which characterise the darker veins of breche marbles generally (see fig. 23).

As the veining extends towards the paler areas, the lines should be made fine and weaker in strength, and the enclosed shapes should be much larger. The colour continues to vary from grey-violet to reddish-violet until the work is completed and lightly softened (see plates 12 and 14).

Several of the strong units are then washed over or partly covered with dilute white, others are coated with Indian red reduced with grey. A few touches of solid white may be introduced here and there amongst the stony parts, with smaller touches upon the dark veins. The latter are so broken up as to resemble miniature pieces of the same species of marble, and it is therefore necessary to employ clean rag over a veining horn to wipe out numerous small stones, some of which may be coloured during the glazing process.

When dry, the work is much improved by glazing with thin washes of the various colours toned down with grey-violet, on the lines described for sienna. Any additions to the whites can now be made, and finally, the fine dark lines are introduced amongst the principal veins.

Chapter Seventeen

Black and gold, vert de mer, vert antique, St. Anne's

Black and gold

This marble is also known as 'Porter', a name which was in all probability derived from Porto Venere, in the Apennines, where it was originally quarried. It is quite distinctive in character, having a black ground clouded with grey, and strongly veined with yellowish, chain-like formations which run fairly parallel (refer to plates 11 and 15).

The palette required for working this marble should include black, white, chrome, raw, and burnt sienna, each pigment being semi-mixed with the 'gilp' so that it may be taken up freely and easily with a large sable pencil. A dipper containing the oil medium will also be necessary.

It is advisable to oil the ground sparingly before commencing with the veining, and to delay the clouding until the main lines have been painted in. The latter may give an appearance of directness and simplicity which is most misleading: it is true that the shapes produced are mainly accidental, but their grouping calls for considerable skill and experience which can be gained only by working from a good example of the real marble.

It is not difficult to judge the correct consistency of the paint. The brush is lightly charged with oil medium and then employed to mix several semi-opaque tints, ranging from pale to warm yellow. By taking up the lighter tone with one side and a warmer colour with the other side of the brush, it is possible, by rolling the side of the pencil across the panel, to form a series of irregular veins of varying length, breadth, and colour. These are linked by fine lines and continued in a meandering but mainly straight direction down the full length of the panel.

There should be continual change of colour, form, and arrangement, some veins being bold and closely massed, with others extended and extremely fine. The general sense of direction is also to be seen in the individual veins which are mainly elongated in form and with smoothly rounded inside shapes, elliptical rather than circular in character. These resemble groups of rounded black stones which are one of the prominent features of the marble.

Other chain-like formations may be put in so as to show about 150 mm of plain background between each. The formations should converge at times and then run more widely apart, but never linking up.

The next operation is that of clouding the background with a dilute wash of dark grey prepared from black and white. A round fitch is particularly useful for this job, and especially for putting in the darker greys within the small stone-like shapes. Very little softening is necessary, and this is mainly confined to the larger areas with perhaps a light stipple upon the smaller ones.

To complete the work, it is necessary to run a few white veins at various angles across the panel. These are unrelated to the principal veins, and may commence almost anywhere and finish by a gradual thinning and fading out.

Vert de mer or black and green

Although this marble differs from the types already considered, it can hardly be described as distinctive; indeed, to the layman, it would appear closely similar to vert d'Egypt and verde di Genova, which are also quarried in Northern Italy, and to Tinos, from the Grecian island of the same name. All are represented by the same process, but the marbler should make himself familiar with the slight differences in form and colour (see plates 12 and 16).

The colours used in painting vert de mer are white, black, and several tints and shades of green. The ground is oiled in, and then scumbled in all directions with a feather charged with dark green, made from raw sienna and Prussian blue, and well diluted with the oil medium (see fig. 22).

At this stage, parts of the work are lightly spotted with the same colour, applied with the tip of an old camel-hair mop. This tool, when stippled upon the palette board, forms numerous points which are equally useful for the production of extremely fine lines.

Stage two consists of veining in a rather indefinite manner with a lighter and brighter tone obtained by the addition of Brunswick green and a little white. The veining may be put in with a feather or with the edge of a large writing pencil, but the effect should be quiet, slightly varied in tone and colour, and placed so as to leave some intervening spaces to be crossed with fine sketchy lines of the same colour.

The principal veins are put in with white, broken with green and a touch of black. These are painted to run across the more crowded masses of under-veining and must be clearly defined and broken in a manner consistent with the real marble. A few veins of pure white may be added, after which the tinted veins are broken by spotting with dilute black.

When dry, the work is glazed with a transparent wash of dark yellowish-green prepared from black, raw sienna, and Prussian blue. Parts of the more prominent veins are shaded and strengthened with the original colours, and any further spotting may be introduced.

Verde antico or vert antique

This beautiful green marble is a brecciated serpentine characterised by various-sized patches of white, black, and dark green, interspersed with a network of streaks in varying tones of green. It was being quarried at Thessaly in Greece, more than 2,000 years ago; the source was then lost, and only rediscovered less than a century ago. It is produced in three varieties – light, moss green, and dark (see plates 12 and 16).

The ground for vert antique may be of black, dark grey, or white, and whilst our illustration worked upon the first named, it is sometimes desirable – especially when suggesting inlay – to employ white, darkening this with a water scumble of black, and then proceeding to work in oil colour.

It is advisable to commence operations by coating the whole surface with a transparent wash – sparingly applied – of middle Brunswick green toned with black. Then, with the ragged tails of a camel-hair mop, or with the tips of a bunch of feathers, the first series of lines are painted in mid-Brunswick green diluted with sufficient oil medium to ensure ease of application and the right degree of transparency.

Upon this we apply a paler series of veins, using the green slightly reduced with white, but still of a translucent nature. Again, and with more white added, the under-veins are streaked and broken with the brighter and more opaque colour, at the same time taking care to leave a number of small patches untouched. The work is lightly softened with the badger before proceeding to add the numerous stone-like shapes.

The latter are mainly angular in form, clearly defined and widely varied in size. It will be noticed that the largest stones are those of the grey or almost black shades, and that the white patches are small, translucent, and frequently present in grouped formations. The smallest stones appear to be spotted upon and enclosed within the meshwork of green veining. Some are dense black, some are more translucent, others grey; all differ in shape, size, and in sharpness of definition.

It may be necessary to do some wiping out with rag moistened in turpentine in order to sharpen the edges of the larger stones or to alter their formation or to increase their number. This wiping out must not be overdone, i.e., it should not result in the complete removal of all lines crossing the area concerned. By allowing some veining to remain, it will be seen somewhat faintly through the superimposed glazed colour, thus securing greater depth and interest.

When the work has become properly set we proceed with the painting of the small black stones, then with the scumbling and edging of the larger pieces. The colours employed should be slightly varied to produce thin washes of black, greenish-black, and dark grey: these shades may be used separately or may – where space permits – be applied conjointly to obtain a cloudy appearance. A one-stroke sable pencil is ideal for the painting of large pieces and the spotting brush is useful for smaller details.

The white calcite fragments are next applied with the sable pencil and colour varying from semi-transparent to solid white, at times slightly tinted with yellow. These, when sufficiently set, are again coated with thin white. In the meantime a fine pencil charged with pale green may be used to emphasise any weak points in the veining, and for the application of the fine veins which run across and around parts of the larger stones.

When the work is quite hard and dry one is able to form an opinion as to the extent of retouching or glazing required. This marble displays considerable tone contrast, but with a sense of repose and flatness; if our painted example is lacking in these qualities we must then secure the necessary translucency and unity by glazing and thereby toning down the harsh places with a very dilute wash containing black and green. Any additions to the central parts of the white fragments can be put in at this stage.

St Anne's Marble

Although lacking the bright colour of other marbles, St Anne's is, nevertheless, very useful as a foil or contrast to the more decorative types. The best variety is quarried in the province of Hainaut, Belgium, and is readily distinguished by its cloudy grey background and the closely packed, greyish-white veins, which form an all-over pattern of loosely connected lines, rather like those found in black and gold marble (see plates 11 and 15).

The painting of this marble is usually carried out upon a black ground, with white as the only colouring agent.

We commence by oiling the ground with a rag moistened in the oil medium, and then proceed to cloud and texture the whole surface, a job which is more easily and quickly completed by applying the dilute white with rag rather than by brushing. Upon this the delicate fossilised shell forms are painted in thin white and the whole is lightly softened.

The principal veins are applied in the manner suggested for black and gold – i.e., by a rolling and dragging motion with the side of a pencil (see fig. 23). In St Anne's marble the veins are more broken and eccentric in form, with numerous enclosed shapes giving a chain-like impression to parts of the work. There is also considerable variation in tone, some parts of the veining being fairly opaque, other parts being translucent, and thereby of a greyish tone.

It is more convenient to employ a small pencil for painting the fine subsidiary veins which run in all directions. It should, however, be realised that a noticeable sense of direction is to be found in this marble and must naturally be shown in the finished work.

Finally, some of the principal veins may be accentuated by edging with dark grey.

Chapter Eighteen

Rouge Roi, granites, porphyries and Cipollino

Rouge Roi

This is a red fossiliferous marble, one of several varieties which are quarried in Belgium and all of which are ordinarily known as 'Rouge'. Its colour varies from a rich red-brown to deep fawn, the whole being broken by irregular patches of bluish-grey and white, with a few prominent veins of opaque white. The mass displays a semi-transparency which gives an effect of fluidity and softness.

The ground for this marble is usually of light grey or white, according to the method of reproduction employed. There is, however, an alternative which combines the advantages of both, with the equally valuable qualities of depth and variegated colouring. It is this alternative which forms the groundwork of our illustrated panel and which we must now describe (plates 11 and 17).

The effect is secured by the application of a blue-grey water glaze, with a little fuller's earth as the binder; this is spread upon a white ground and immediately stippled with a coarse sponge and softened with the 'badger'. Some parts are allowed to remain spotty, while other areas are softened until variations of tone are just discernible. Then, with a damp wash-leather and veining horn, parts of the watercolour must be wiped out to represent the fossils and principal veins: the work dries rapidly and is then continued in oil colour.

In the meantime we arrange the appropriate colours upon a palette, prepare the oil medium in one dipper and turpentine in another. Several paint-brushes, fitches, and sable pencils are required, as well as clean rag, paper, and hog's-hair softener, and the camel-hair 'spotting' brush. The dry surface is then moistened with a little oil medium in readiness for the application of oil glazes.

The 'rubbing-in' colour is prepared from Venetian red and ochre, diluted with oil medium until of a translucent nature. This is applied upon the grey parts only, leaving the white veins and a number of grey patches untouched. The darker reds must now be introduced, some parts of the work being strengthened with burnt sienna, others being coated with Indian red, toned with black, the whole being completed without any encroachment upon the grey or white areas (see plates 11 and 17).

It is now necessary to break up the masses of red by uncovering parts of the grey and white underwork, but to achieve the desired variation in shapes and sizes we employ both the spotting brush and a piece of crumpled paper, dipped in turpentine and then in blue-black (a mixture of black and ultramarine blue) for the stippling of parts of the surface. This produces an infinite variety of tints and shades of grey, and also helps to impart a stony appearance to the work. It is important, whilst 'spotting' or breaking the ground, to control the solvent action of the turpentine by dabbing the stippled work with a soft rag. This prevents the formation of excessively large patches of grey, and also checks any tendency to run.

Some of the grey patches may now be touched here and there with a wash of thin white; other parts are glazed with varying tones of grey, and several of the larger patches may be outlined with both grey and white. The whole is now softened with the 'badger', after which any further 'spotting' may be done with reds and greys. The work is now complete except for the strengthening of the white veins; these are coated with dilute white, broken with a touch of ochre.

When dry, the reds may be toned down, and any necessary unity secured by glazing with dilute blue-black; some fine lines may also be put in with Indian red.

St Rémi

This marble, so-called because it is quarried at St Rémi, on the frontier of Luxembourg, is worthy of mention on account of its distinctive markings. It is closely related to the Rouge type and is painted in exactly the same manner as Rouge Roi, but with larger masses of grey, which give a striking display of intricate and curly shapes.

Granites and porphyries

These granular substances do not correctly rank as true marble, yet they take a good polish, are extremely durable, and are much used in marble decoration. In appearance they are solid, substantial, and cover an exceedingly wide range of interesting broken-colour effects. They are universally spread over the earth's surface, each country producing several distinct varieties.

Granites quarried in the British Isles include the red Peterhead granite, the grey Aberdeen granite, and varieties of pink, blue-grey, and light greenish-grey. Some present a coarse and mottled appearance, while others are fine and spotted. The porphyries include almost every conceivable colour, thus extending our field of broken-colour suggestions.

All may be represented in paint by the same straightforward procedure of sponge stippling; the darker tones being applied first and the paler or predominant colours last. A translucent appearance is quite as valuable in this work as in marbling, and to achieve any measure of success it is necessary to employ a pale background and to make the most of its luminosity by superimposing oil colours of thin consistency.

Peterhead granite

This is worked upon a white ground by the following methods:

First, 'rub-in' with a mixture of white and burnt ochre, matching as nearly as possible the middle tint of the granite. Whilst this is setting, prepare small quantities of (a) black; (b) bluish-grey; (c) white; (d) red, to match the darkest tone in the real granite – made from Venetian red and burnt umber. Several sponges of a coarse, open type will be required, and also a large palette board.

All being ready, a sponge is dipped into the black, dabbed upon the palette to remove superfluous colour, and then stippled lightly over the work already 'rubbed-in', until the surface is mottled as uniformly as possible. The grey is applied in the same manner, and then the white, each with a separate sponge.

The dark red, which forms the dominant colour of the granite, should be more opaque in order to produce a surface stipple which, in consequence of its solidity, intensifies the transparency of the underwork. Although the various colours are applied in a uniform manner, sponge stippling should ensure just the right balance of irregularity of shapes and sizes as seen in the natural substance. Care and judgment are necessary to secure the correct distribution of each colour, but this is quickly ascertained by practice (see plate 11).

There is another process which is sometimes employed in the painting of granite and porphyries, but owing to the minute size of the spots produced, and to their uncertain distribution, any process involving the application of colour by spattering or splashing must obviously be limited in its uses. Furthermore, there is a feeling of harshness and a lack of translucency about spatter work which militates against its general use in marbling.

Cipollino

The decorative value of this marble was appreciated in ancient Rome, where its laminated green and white formation provided striking colour contrast to the more neutral backgrounds of many public buildings. Some idea of its widespread use can be gathered from the number of columns and other architectural features which may still be seen – often in excellent condition – in and around this centre of the Roman Empire.

In London, we are fortunate in having a magnificent display of both the Greek and Swiss varieties of Cipollino – in addition to other marbles mentioned in this volume – which may be seen under excellent conditions of lighting in Westminster Cathedral. Here, in a perfect and noble setting, is material which cannot fail to inspire the painter and colourist.

The name Cipollino is applied to marbles which have a whitish ground traversed with veins or bands of green talc. A grey variety, called Cardiglio, is quarried in Sicily, and another type, Apollazzo, is white, streaked with violet. The pattern displayed upon opposite sides of a circular pillar or column resembles an eccentric and curly example of the heart-grain of timber, but the intervening spaces are similar to the normal cross-cut formation as seen in flat slabs of the marble (refer to plate 12).

To represent this marble the white ground is moistened with oil medium slightly tinted with grey-green (obtained by mixing raw sienna and Prussian blue). Upon this we sketch a series of bands of stronger grey-green, varying tone and colour by the introduction of touches of mid-Brunswick green. Colours should be semi-transparent and applied sparingly, using the side of a flat fitch, but keeping the light veins clean, sharp, varied in thickness, and of a slightly undulating character. Soften in the general direction of the strata.

When the work is partly set, we put in the fine dark lines with grey-green toned with black, then, with a larger pencil, add several flicks of thin white upon the more wavy portions of the light veins. The whole is finally softened and stippled lightly.

Chapter Nineteen

Other broken colour effects and varnishing

At the time of revising this book a vast range of broken colour effects in conjunction with graining and marbling is enjoying a resurgence of popularity in the decoration of historic, commercial and domestic buildings.

These finishes use many of the techniques and materials of graining and marbling. Oil scumble glaze is designed to 'hold-up' after being textured rather like oil graining colours. Other effects require the more liquid medium 'gilp' which is required when producing marble effects. The grounds for the following techniques are the same material as for graining and marbling.

Clear oil scumble glaze

This material is available ready for use requiring only the addition of white spirit and the appropriate stainers. The formulation of oil scumble glaze allows time for its application and subsequent texturing before it 'sets-up'. Linseed oil may be added to extend this time on large areas, but it must be remembered that while still wet this material is very vulnerable to damage which normally cannot be successfully touched in. To colour the clear glaze, artists' oil colours or pigments bound in oil are used. Universal stainers (the type used for tinting both oil and water paints) will tint the glaze but tend to be rather opaque and slow drying. To ensure good distribution of colour throughout the glaze the stainers should be broken down with a small amount of the glaze on a palette board before mixing it into the bulk of the material. This should be added slowly to the clear glaze to allow for colour checks and prevent over staining. After staining, the material is strained and further thinners added if required. The addition of thinners will lighten the finished affect and soften the texture produced.

Plate 13

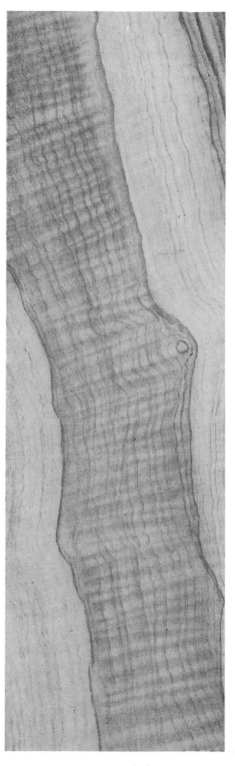

Walnut

Italian walnut

Plate 18

Hair stipple

Rag roll or rolling

Dragging

Plate 18
(Cont.)

Shaded or blended

Glaze and wipe or distressing

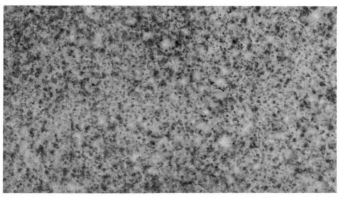

Spatter

Plate 18
(Cont.)

Tortoiseshell

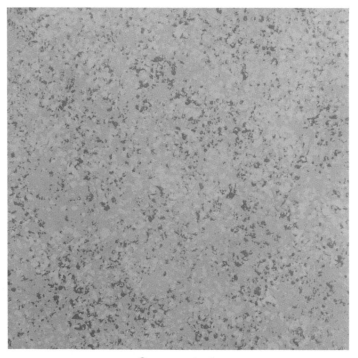

Sponge stipple

Stippling

This is the standard technique when using oil scumble glaze both as a finish and as an essential part of other effects. After brush application the surface is firmly stippled with a pure bristle hair stippler to remove the brush marks and even out the tinted glaze. A second lighter stippling is then required which will produce the delicate two-tone effect of the ground colour and the superimposed tinted glaze. When dry which is usually after about forty-eight hours, the work should be protected with at least one coat of oil varnish.

Ragging or rag rolling

This popular effect produces a texture similar to that of crushed velvet, the subtlety of which depends on the thickness of the glaze used and the closeness of tone between ground colour and tinted glaze. The glaze is applied and stippled and, while still wet, a large piece of crumpled rag (preferably lint free cotton) is rolled through it. To achieve an even texture the rolling should be carried out in all directions slightly overlapping each pass of the cloth. Because the pressure employed on the brush or rag affects the finished texture, the principle of only one operative carrying out the manipulation technique is most important.

Dragging

This is a texture similar to that of brushgraining which is popular both as a finish for furniture or large wall surfaces. This effect does present problems on poorly prepared surfaces and those containing architectural features which impede the movement of the brush.

The coloured glaze is applied by brush and then stippled to obtain even weight of colour. A clean loosly filled brush such as a flogger, dusting brush or, on large areas, a paperhanger's or distemper brush, is drawn firmly through the glaze to produce the fine silky irregular striped effect. The various problems that can mar this texture are as follows:

(1) The glaze will form in dark blotches in indentations on surfaces of poor preparation.
(2) The brush must be drawn through the glaze without interruptions. Stopping half-way will result in a broken striped effect. This problem

is exacerbated on high walls when stepladders have to be used as there is a tendency for the brush to drift away from the vertical when the operative climbs down the angled steps and at the same time drags the brush through the glaze.

(3) When the dragging brush comes into contact with a break in the surface such as picture rails, skirtings or carved mouldings, the brush must be lifted, which will leave small areas not dragged which can, if extensive, spoil the overall effect. Only experiment and practice by the decorator can solve or lessen this problem and in some cases the only solution would be to select an alternative effect.

Shading or blending

This subdued effect of a colour softening into a lighter tone can be used on large flat areas such as walls, but is at its most effective when highlighting the inside of wall and door panels.

The various tones of glaze are applied in bands merging one into another initially with a brush. The darkest tone is applied inside the moulding, gradually getting lighter towards the centre. The work is then stippled firmly from the centre towards the darker glaze around the moulding, making sure that one side of the stippler faces outwards at all times.

This initial texturing will blend the tones into each other creating a graduated effect. A final light stippling from the moulding towards the centre completes the effect.

Glazing and wiping: sometimes termed 'distressing'

This technique is very useful to highlight and enhance enrichments on carved or moulded surfaces.

The required coloured glaze is brushed over and into the relief decoration and evened out using a hair stippler, stencil brush or the tips of a worn paintbrush depending on the nature or size of the work in hand. A flat pad of cotton rag is then wiped over the surface exposing the original ground colour on the highlights of the detail leaving the tinted glaze in the surface indentations.

All the foregoing techniques in some way expose the ground colour and so the colour of the resulting effect will be a combination of the ground and the tinted glaze, a yellow ground superimposed with a blue glaze will appear very green. Bright red glaze will appear pink on a white

Plate 14

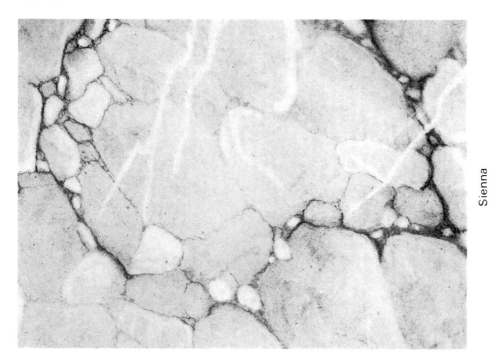

Sienna

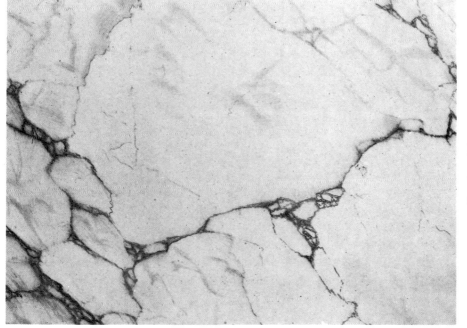

Breche violette

Plate 15

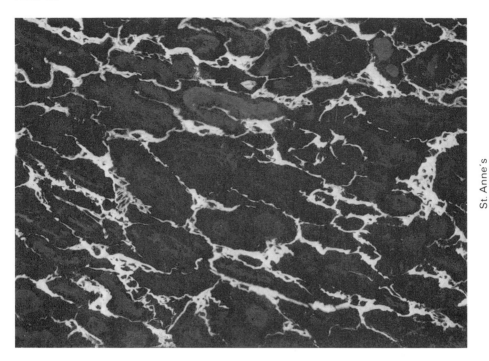

St. Anne's

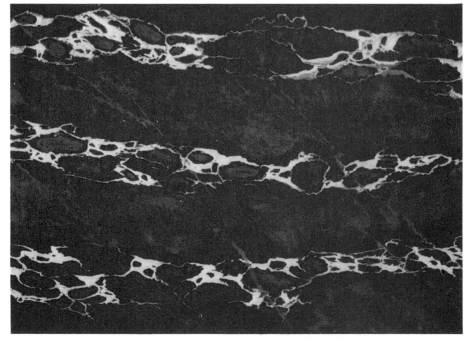

Black and gold

Plate 16

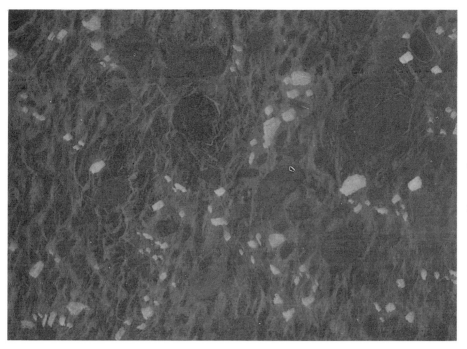

Vert antique

Vert de mer

Plate 17

White vein

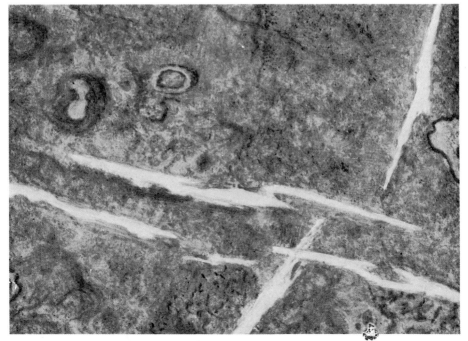

Rouge Roi

ground. For this reason sample panels should be prepared before the ground colours are applied.

The extensive open-time of oil scumble glaze enables many interesting effects to be produced. Experiments using crushed paper, plastic and rubber combs can be used to produce a great variety of textures and patterns. The use of a range of colours on one surface and then rag-rolled will produce interesting marble-like effects.

Spattering

This effect is best executed with the 'gilp' medium used for marbling. Oil colours are mixed with the gilp into a very thin coloured glaze, which can be brushed or flicked over the surface in as many colours as desired. This in its turn can be 'opened-up' to expose the ground colour by spotting with white spirit or turpentine using a small brush or a piece of natural sponge. The range of textures, patterns and colourations are only as varied as the imagination of the decorator. When dry the work should be varnished to provide a hard durable finish.

The main problem encountered with this technique is the tendency for the work to sag or run before it dries, making it more difficult to carry out on vertical surfaces.

Tortoiseshell and similar effects can be achieved by blending patches of thinned tinted gloss varnish with a badger softener.

Very sparkling effects are produced by spattering on metallic paint or foil grounds.

Sponge stipple

This multicoloured speckled texture is possibly the easiest to produce and suitable for almost any size or type of surface.

All that is required is a large natural sponge and a roller tray or flat dish to hold the paint. Thinned oil paint, tinted oil scumble glaze or emulsion paint are all suitable for producing this effect, the advantage of emulsion paint being its speed of drying. The technique consists of dipping the tips of the sponge into the colour and printing it onto the surface. When one colour has dried subsequent colours may be added. Greater durability can be achieved by varnishing.

Varnishing

The materials used for graining, marbling, and other broken colour effects using scumble glaze have little protective qualities, and are not very resistant to abrasion. For this reason it is essential to protect the work with clear, pale varnish.

Varnishing also enables a suitable degree of sheen to be selected for the appropriate work. The dull gloss of marble and waxed wood is best imitated with eggshell finish varnish, while the high gloss of French polished timber is simulated with full gloss varnish. Matt varnish is also available for surfaces requiring no sheen at all.

Although polyurethane varnishes are available, it has been found that traditional oil, or oil/alkyd varnish is superior. It maintains its sheen and protective qualities longer, and is highly recommended for exterior and interior work. Eggshell and matt varnishes have limited protective qualities, and are suitable only for interior work.

The protective qualities of eggshell and matt varnishes can be increased by first varnishing with gloss varnish, and after several days hardening, wet abrade with grade 400 silicon carbide abrasive paper and refinish with eggshell or matt varnish. It is the glossy undercoat which forms the real protective film, and which enhances the richness and depth of the work.

Very high quality gloss varnish work can be achieved by building the surface with two or three coats of gloss varnish and allowing to harden for several days before wet abrading, and applying the finishing coat of varnish.

Applying the varnish

The success of varnishing, and indeed painting depends on the scrupulous cleanliness of the surface to be coated, the equipment and material used, and even the atmosphere. By following as many of the precautions listed below, a high quality finish should be obtained.

Material

(1) Stand material at room temperature for at least a day.
(2) Dust off top of container before opening.
(3) Follow manufacturer's instructions on stirring. Some materials,

especially eggshell and flat varnishes, require thorough stirring, while some gloss varnishes do not.
(4) Strain through a fine filter into a clean pot.

Brushes

(1) Keep a separate set of brushes for varnishing and painting.
(2) Wash brushes in detergent, rinse in clean water and allow to dry.
(3) When varnishing, work brushes into a small amount of varnish to eliminate frothing and milkiness, and then discard the contaminated material.

Surface

(1) If possible, raise the item to be coated off the work bench or floor.
(2) Remove surface dust with an adhesive duster (Tac Rag) immediately before application.

Environment

(1) Ensure all surfaces are clean, and all room openings are closed.
(2) Enter the room and work without disturbing too much air.
(3) Ensure normal room temperature during application, and until coating is dry.
(4) Prevent cold air draughts across the surface during drying, as this can cause some varnishes to dry with an opaque finish.

Application

(1) Avoid wearing loose fibred clothing during application.
(2) Follow manufacturer's instructions.
(3) Ensure a full, even coat with no misses or scuffed areas.
(4) Apply quickly while maintaining a 'wet edge'.
(5) Do not return to areas previously coated.

Glossary of terms

Blending: A process of merging different coloured scumble glazes together on a prepared ground coat with no apparent joint between the colours.

Broken colour: Decorative effects produced by manipulating coloured translucent media over a harmoniously coloured ground coat. For example, graining, marbling, rag rolling, dragging, etc.

Brush graining: A process of imitating a simple straight grained effect by dragging a dry brush across a surface previously coated with graining colour.

Burr wood: The curly grain pattern of knots or burrs displayed on woods cut from the root section of the tree, such as burr walnut.

Check rolling: A process of imitating the broken pore marks found in woods such as oak, by rolling a serrated check roller charged with graining colour over the grained surface.

Cissing: Failure of a coating to form a continuous film on the surface. For example, water graining colour rolls back, or cisses into globules if applied over a greasy surface.

Clouding: Often the first process of marbling by applying transparent colours followed by scumbling or rag rolling and finally softening and blending.

Combing: A process of drawing combs of steel, or other more flexible material across a surface previously coated with graining colour or coloured scumble glaze.

Dapples: Another term for 'silver grain'.

Dappling: A process of imitating the silver grain, or dapples found in woods such as quartered oak, by wiping-out, or by painting-in the pattern.

Darks: See 'Lights and darks'.

Dashes: Another term for 'silver grain'.

Degreasing: The process of removing grease from a surface before applying water graining colour. Failure to degrease will result in cissing of the graining colour.

Distressing: Another term for 'glazing and wiping'.

Dragging: A broken colour effect produced by dragging a dry brush across a surface previously coated with coloured scumble glaze.

Feather grain: The feather grain pattern displayed on woods such as feather mahogany, cut from the upper part of the tree at the intersection of two main branches.

Figuring: The process of imitating the main design, or pattern of a wood grain as opposed to a simple straight grained effect.

Fixing: Another term for 'strapping down'.

Flogging: A process of imitating the fine pores of wood by beating a dry brush against a surface previously coated with graining colour.

Gilp: A term used for the oil medium for marbling. Consists of two (2) parts of turpentine and one (1) part of raw linseed oil plus a small amount of liquid driers.

Glaze coat: Graining medium coloured very sparingly with pigment so as to remain very translucent. Used to subdue strongly coloured or patterned work.

Glazing and wiping: A broken colour effect produced by applying coloured scumble over and into the relief of carved or moulded decoration, and wiping the highlights with clean rag to expose the ground colour.

Graining: The process of imitating the simple straight grain of a wood as opposed to the more decorative pattern of figured wood usually displayed alongside.

Graining colour: Graining medium containing a suitable pigment to give it colour. Applied sparingly to a harmonising ground colour and manipulated to imitate the grain pattern of wood.

Graining medium: The clear liquid part of graining colour which requires to be coloured with added pigment. Both oil and water media contain a binder, thinner, and when necessary, 'megilp'.

Ground colour: A suitably coloured hard drying paint of medium sheen upon which graining, marbling, and other broken colour effects are carried out.

Heartwood: The elliptical grain pattern displayed on woods obtained by sawing through the length of the tree and cutting through many annular growth rings.

Hold-up: The quality of the medium which after manipulation with brushes or combs prevents the pattern flowing together.

Lights and darks: The patterned grain of wood may be seen as a dark pattern on a light background, or a light pattern on a dark background. These contrasting areas of colour are known as 'lights' and 'darks'.

Liquid driers: Added in small quantities to oil graining colour to ensure overnight drying. Terebine driers and patent driers are two types commonly available.

Medullary rays: Rays of sap which radiate out from the centre of the tree. Quartered wood such as quartered oak displays this as silver grain or dapples.

Megilp: Added to graining colour to ensure that the graining 'holds up' and does not flow together. Scumble glaze is commonly added to oil graining colour as a megilp.

Mottling: The process of imitating the soft lights and darks seen beneath the surface grain of many woods, by removing areas of graining colour and softening.

Oiling-in: The process of lightly moistening a ground colour with oil medium before marbling. Oil colours are more easily applied to an oily surface than a dry one.

Opening-up: The process of breaking-up the colour on the surface by dabbing with a natural sponge moistened with turpentine, and softening.

Overgraining: The process of adding greater depth to graining work by painting fine grain lines, mottling, flogging and stippling.

Painting-in: A method of producing figured graining by painting the pattern with darker graining colour, as opposed to 'wiping out' the pattern.

Pigment: Finely ground coloured powders added to graining medium to give it colour. Obtained ready ground in oil and known as oil stainer, or in dry powder form.

Quartered wood: Wood cut from a tree which has first been sawn lengthways into four quarters. In oak this displays the 'medullary rays' known as silver grain.

Rag rolling: A broken colour effect produced by rolling crumpled rag or paper across a surface previously coated with coloured scumble glaze.

Root wood: The curly grain pattern displayed on woods cut from the root section of the tree, such as root of oak.

Rubbing-in: The process of applying a thin, well rubbed out coat of graining colour, or coloured glaze to a previously applied and hardened ground colour.

Sapwood: Another erroneous term for 'heartwood'.

Scumble glaze: Clear or ready coloured oil medium which is manipulated to produce broken colour effects. Also added to oil graining medium as a 'megilp'.

Scumbling: The process of producing a broken-colour effect by manipulating a coloured scumble glaze over a harmoniously coloured ground coat.

Setting-up: The stage at which graining colour and coloured scumble glaze begins to dry and is no longer workable on the surface.

Shading: The process of blending a coloured scumble glaze from one tone of colour to a lighter tone with no apparent joints between the tones.

Sharp coat: Material such as graining colour, scumble glaze, paint or varnish considerably thinned by the addition of a suitable solvent.

Silver grain: The flowing broken grain pattern of medullary rays displayed in many quarter sawn woods, such as quartered oak. Also known as dapples or dashes.

Softening: Various processes of fading-out the sharp edges of patterns produced when manipulating the materials used for graining and marbling.

Spattering: A broken colour effect produced by flicking or spraying seperate specks of colour onto a surface. Also used in marbling and overgraining.

Sponging: A broken colour effect produced by either applying colour or removing colour by dabbing with a natural sponge. Also the process of 'opening up' marbling.

Spotting: A process of producing round openings in wet marbling by flicking specks of a suitable solvent onto the surface and softening.

Stippling: The process of removing the brushmarks from, or simply texturing a wet coating by dabbing with brush, rag, sponge or paper.

Straight graining: The process of imitating the straight grain of wood, as opposed to the more decorative figuring of heartwood, burr wood and quartered wood, etc.

Strapping down: The process of fixing graining carried out in water media by applying a 'sharp' coat of varnish. This enables further coatings to be applied without disturbing the underlying graining.

Tone: The lightness or darkness of a colour. The tone of graining colour and scumble glaze is usually only slightly different to that of the ground colour upon which it is applied.

Veining: The process of imitating the pattern of veins and fine lines of marble.

Wiping out: A process of producing figured graining and veining by removing colour with rag and veining horn, as opposed to 'painting-in' with colour.

Index